£14.90

20 1178658 6

TELEPEN

LEICESTER UNIVERSITY LIBRARY

below.

AF795570

THE SCOTTISH SALT INDUSTRY

Gable wall of panhouse, Cock, Arran. Constructed *c.* 1710 by the ducal Hamilton family. The works operated intermittently until the 1730s.

THE SCOTTISH SALT INDUSTRY 1570 - 1850

an economic and social history

Christopher A Whatley

ABERDEEN UNIVERSITY PRESS

First published 1987
Aberdeen University Press
A member of the Pergamon Group

© Chris A Whatley 1987

All rights reserved. No part of this publication may be reproduced, stored in a retrieval system, or transmitted in any form or by any means, electronic, mechanical, photocopying, recording or otherwise, without the prior permission of Aberdeen University Press.

British Library Cataloguing in Publication Data

Whatley, Christopher A
The Scottish salt industry, 1570–1850 : an economic and social history.
1. Salt industry and trade—Scotland—History
I. Title
338.4′766142′0944 HD9213.G7353

ISBN 0 08 034528X

LEICESTER UNIVERSITY LIBRARY

D41
WHA
2011786586

Typeset on a Lasercomp at Oxford University Computing Service

PRINTED IN GREAT BRITAIN
THE UNIVERSITY PRESS
ABERDEEN

CONTENTS

ILLUSTRATIONS

MAPS

FIGURES

TABLES

PREFACE AND ACKNOWLEDGEMENTS

Inevitably, there are several references in this book to quantities of salt. Weights and measures in Scotland however, even in the nineteenth century, were far from standardised. As far as salt is concerned, the problem is far more severe prior to 1713, when duties imposed by the English parliament during the negotiations which preceded the Union of 1707 began to be collected. Post-1713 salt taxes were paid by the Imperial bushel of 56 pounds (36.268 l.) and accordingly, legally manufactured salt was weighed by the bushel. Some of the difficulties faced by the aspiring quantifier for the pre-1713 period are outlined in the Appendix. None of the pre-1713 quantities should be considered as anything other than approximate indicators. The chalder of salt may have been around 60 bushels in Fife, but elsewhere its precise weight cannot be ascertained at present; bolls were usually either 'large' or 'small', that is just over four and three bushels respectively; the ton was nominally 40 bushels; the wey was around a ton of 2,240 pounds. Even if the measures were certain, there could be no certainty about the accuracy and honesty of the quantities recorded by the measurers, the shipping clerks, the salt grieves and the customs and excise officers.

As most costs and prices in the book are taken from pre-1972 sources they are quoted in £*s.d.* currency. For any younger reader who is unfamiliar with these values, and who wishes to convert them into decimal currency figures, it may be helpful to know that 2*d.* = 1p and 1*s.* = 5p. A further complication is that many sums quoted, especially those pre-*c.*1750, are given in £ Scots, which around 1700 was worth one-twelth of the English £ sterling. Where Scottish currency values are used, this is stated in the text. Dates have been modernized to begin the year on 1 January.

Lack of space has prevented the inclusion of a separate bibliography, but it is hoped that this omission will be largely overcome by the extensive set of Notes which include most of the sources which were used in the preparation of this monograph. They also elaborate upon some of the points made in the text, detailed discussion of which might not be of interest to the general reader.

In the preparation of this book I have been assisted by numerous people. In naming several of them here, I trust that those who are not listed individually will accept my sincere thanks for the help they have given me. I was only able to undertake the necessary research after being appointed as a Lecturer in Economic History in the Department of Modern History, University of Dundee. For their encouragement during the time it has taken to produce this volume I wish to express my sincere gratitude to my former colleague and head of department, Professor Peter J Parish, and his successor Professor C J Bartlett. Annette Smith's enthusiasm for this project and her willingness to lend me books and to bend a sympathetic ear, provided inspiration during those times when my own commitment was weakening.

For their practical help, my debt to the staff of the National Library of Scotland's South Reading Room is enormous. Their courteousness and the promptness with which my constant requests for documents were met never ceased to impress. The staffs of the Scottish Record Office and the Signet Library also provided excellent service. Many local librarians and archivists were extraordinarily helpful: Isabel Walker of Hamilton Reference Library; Murdo Macdonald, archivist, Argyll and Bute District Council; Walter Makey, formerly archivist, Edinburgh District Council; Alison Fraser, archivist, Orkney; Robert Smart, archivist, University of St Andrews; Gordon Booth, Islay Museums Trust; Richard Dell and Andrew Jackson, Strathclyde Regional Archives; G H Ballantyne, of the Signet Library, Edinburgh and Christopher Neale of Dunfermline Central Library. I am grateful too to the staff of the libraries of the Universities of Aberdeen, Dundee, Edinburgh and St Andrews. Extremely helpful in assisting me to find illustrative material were Jim Wood of the Royal Scottish Museum, Geoffrey Stell of the Royal Commission on the Ancient and Historical Monuments of Scotland, Raymond Lamb of the Orkney Heritage Society, Rosalind Marshall of the Scottish National Portrait Gallery and Angela West of the School of Scottish Studies, University of Edinburgh. Requests for information about holdings and potential sources were courteously treated by Michael Jubb of the Public Record Office and the staffs of Chester Record Office and Northumberland County Record Office.

I have had the good fortune to be able to draw on much invaluable archive material which is held in private hands. For allowing me to have access to this I wish to express my appreciation to His Grace the Duke of Hamilton, who permitted me to use his collection at Lennoxlove, some of which is now in the Scottish Record Office; Rt Hon the Earl of Cawdor and his extremely personable factor, Alan de Candia; the Earl of Dundonald; Mrs Hanbury-Tenison, owner of the Ramsay of Wardlaw MSS; and last, but foremost in terms of hospitality and interest in the subject of my research, Mrs Alice Cunnack, who made it possible for me to examine some of the treasure trove of historical materials held at Wemyss Castle in Fife.

Many other people have freely given of their time in response to my queries and requests for information, advice or guidance, including Graham Cummings, Colin Martin, William Walker, Kenneth Simpson, Graham Douglas, Joyce Ellis, Eric Simpson, Bruce Walker and Angus Calder. Alexander Fenton provided me with some useful references to published work on the salt industry overseas in the early stages of this work. Especially warm thanks are due to John Hume and Ian Donnachie, as well as to David Abercrombie, for his statistical and computational skills, and to Isabel Guy, for letting me have a preview of parts of her now-completed thesis. The mistakes and errors of judgement however are all my own.

Without funding for the necessary travel to libraries and archive centres most historical research would be extremely difficult to undertake.

Accordingly I wish to acknowledge the generous assistance provided by the Carnegie Trust for the Universities of Scotland.

Part of my agreement with Aberdeen University Press was that I should deliver the manuscript in camera-ready form. This would have been extremely difficult without the initial assistance of Professor Rex Last of the Department of Modern Languages, University of Dundee. It would have been impossible without the technical expertise, guidance and labour which have been given so generously by John Bochel, also of the University of Dundee, and his son Hugh, of the University of Kent at Canterbury. Their efforts have been complemented by David Murie and others of the Computing Centre at Dundee. Doris Tindall made valiant efforts to master a word-processing system which she was unfamiliar with while Mrs Jessie Young turned some of my earlier drafts into a readable form. The diagrams and maps were more than competently drawn by Carolyn Bain of the Department of Geography in Dundee.

Lilian, Eilidh and Neil will be glad to see the back of this. My daughter Eilidh is anxious to discover whether her father can write about anything other than saltpans.

University of Dundee

Christopher Whatley

INTRODUCTION

'Great Quarrels', Daniel Defoe observed, broke out during the last weeks of the Scottish Parliament in December 1706, as members discussed the level at which English salt taxes were to be imposed after the proposed parliamentary union had become a reality in 1707. Exploiting popular opposition to a tax which would hit the Scots poor particularly hard, anti-Union forces had hoped that by making the impossible demand that Scotland should be perpetually exempt from the salt tax, which the English Parliament would then have to refuse, the 'Union might split upon that Rock'.[1] That they failed in their attempt to scupper the Union should not be allowed to divert attention from the fact that its opponents had been right to recognise the passionate feelings that the subject of salt and its price could arouse in early modern Scotland. For Lord Belhaven to use the image in his well-known anti-Union address, of Scotsmen being forced to eat 'saltless pottage' under a united British parliament, may well seem laughable to modern eyes, but to his contemporaries it would have been a highly potent form of persuasion. For in Scotland, as in the rest of Europe, salt was then an 'important and necesary' commodity, making survival possible, through the preservation of meat and fish for winter use; and everyday diets tolerable, with its addition to the 'farinaceous gruel' which formed such a large part of the calorific intake of our ancestors until relatively recent times.[2] Man may not have a physiological need to supplement his diet with grains of salt, but all the indications are that he has an almost unquenchable salt appetite.[3]

Yet, strangely, lists of the component parts of the Scottish diet in the period with which this book is concerned often omit salt, or mention it only in passing,[4] even though it was universally used, by all ranks of society. Prior to the middle of the eighteenth century it could be the sole commodity which brought the peasant, as a consumer, within the orbit of the market economy. As the standard of living of the Scottish handloom weavers deteriorated in the early nineteenth century, so salted fish and foodstuffs such as oatmeal and potatoes, which benefited from the addition of salt, formed an increasingly large proportion of their diets.[5] Earlier, James Boswell, on his highland tour with Dr Johnson, noted that the only imports to the Hebridean island of Iona, were iron and salt,[6] although in parts of the west highlands as well as Galloway, it was not unusual to find the inhabitants making their own salt, from sea water, in communal kettles and boilers, or on their own, in cans, at the domestic hearth.[7]

Most Scots, however, used commercially produced salt. Much of this was made in Scotland, using the only raw materials available: sea water, boiled in lead, and later iron, pans (the iron for which had to be imported in the pre-Carron era), by fires of turf, peat and coal. In a European context, this means of making salt was unique, being carried on only in Scotland and along England's north eastern coastline. For Scots consumers however, and the overseas customers of the country's saltworks, prior to the great expansion of the Cheshire rock salt industry in the eighteenth century there was often no alternative. Imports of the preferred 'Bay' salt, necessarily brought by sea from France, Portugal and Spain, were vulnerable to interruption by foul weather, war, and naval blockade. That the native product was otherwise the sole means of maintaining a supply of salt in Scotland, was given by Viscount Stair, in his *Institutions*, as the principal justification for the introduction of what he considered to be a form of slavery amongst the Scottish salt workers from 1606.[8] Initially at least, their servitude was in the national interest. From the 1660s, the Scottish parliament insisted that, for domestic purposes, only Scottish-made salt be used. Lower salt taxes imposed on the native product at the time of the Union of 1707 ensured its status as the country's leading kitchen salt until 1823, used not only in cooking, but also to preserve and season butter, cheese, meat and fish. As the forerunner of the modern Scottish chemical industry, it was also used in a number of manufacturing processes, as a mild abrasive by leather workers, a glaze by potters, and as one of the raw materials used to make bleach.[9]

The Scottish salt industry has not, until now, been the subject of serious investigation. Generally speaking, it has been overshadowed by coal mining. There is some justification for this: by the later eighteenth century salt making was very much an offshoot[10] of the Scottish coal industry, providing a useful outlet for the large quantities of 'small' coal which were brought up at many pits. In the nineteenth century coal provided the raw material base for Scotland's industrialisation. The part played by Scottish salt in this process was peripheral. Intimately linked however, the positions of the two industries had shifted over time. One of the objectives of this book is to suggest that salt had not always been the junior partner. Its relative insignificance by the time of Queen Victoria's reign should not be allowed to obscure the prominent part it played in previous eras. In the early 1600s, with James VI on the throne, exports of Scottish salt had been more valuable than those of coal.[11] Well into the eighteenth century, many of Scotland's coastal collieries remained heavily dependent upon sales of their coal to the saltworks: indeed without the latter they simply could not have survived the major difficulties which began to face leading east coast coalmasters in the late seventeenth century, as tariffs and growing competition closed off formerly lucrative markets overseas. At the same time, following the vast expansion of coal mining which had occurred during the 1600s, and the intensive mining of the most accessible seams, the future for the coal industry appeared far from secure. 'It is as Evident as Sun Shine', wrote one observer around the turn of the eighteenth century, 'that four parts of five of all the Coals of Scotland (which hath Communicatione with the Sea) is exhausted', and that the 'whole Advantage of Scots Coalmasters Depends on their Salt'.[12]

When salt first began to be made by evaporating sea water around the coast of Scotland is unknown. Its beginnings certainly antedate the first documentary indications of large scale salt production, which come from the twelfth century. Saltpans were then in the possession of both the Crown and several religious houses, such as Dunfermline, Holyrood, Kelso, Kilwinning and Newbattle.[13] Its manufacture was clearly judged to be of considerable importance: following the transfer of some pans on the upper reaches of the Forth by King David I to Dunfermline abbey in 1140, the workers were to be 'protected by the king's Peace', while by the end of the following century the office of 'Master of the Royal Saltworks' had been created.[14] Precisely what course the industry followed during the next three centuries is uncertain. What can be said is that by the middle of the sixteenth century, salt manufacturing had become firmly established in many parts of Scotland. Shoreside outcrops of easily mined coal, local demand and increasing opportunities for sales overseas, encouraged concentration on the north and south banks of the river Forth. At this stage of its development, the industry was dominated by the pans which had been built on the short stretch of coastline which lay between Musselburgh and 'Salt Preistown' or 'Prestoun', later to become Prestonpans.[15] Even so, salt making also had a presence on the west coast, not only at Saltcoats, but also further south, in the burgh of Prestwick, at a site which would later be known as Marypans or Craigie.[16]

The 'golden age' of Scottish salt manufacturing, however, began in the 1570s. Built initially on seven decades of remarkable success in export markets, a patchy prosperity was maintained, largely by capturing the home market, until the 1820s. Thereafter the industry underwent a long period of decline, eventually disappearing in 1959, when the remaining saltworks at Prestonpans was closed down.[17] It is partly to describe and account for the industry's fortunes during the two and a half centuries after around 1570 that this book has been written.

Although sea salt is no longer made commercially in Scotland, place names provide lasting testimony to the scale of its former presence in a number of localities. The most obvious are Saltcoats, on the Firth of Clyde, and Grangepans, Kennetpans, Newpans and Prestonpans, on the Forth. The banks of Scotland's eastern trading artery, according to the English visitor Sir William Brereton in the 1630s, were lined with 'innumerable' saltworks. His view would have included Culross and Kincardine: in 1722 the latter town was said to have been 'made up mostly of salt pans and those that are employed in them.' Closer scrutiny of cartographic evidence reveals traces of salt manufacturing in less likely places. Two 'Saltpan Bays' for example are to be found on Galloway's western seaboard, where numerous small peat-fired saltworks were constructed, mainly in the seventeenth and early eighteenth centuries; on Islay the English 'Salt Pan Point' stands out conspicuously between the Gaelic Rubha Buidhe and Rubha an t-Saile on Loch Indaal, virtually the sole surviving site-related trace of a failed effort to establish a saltwork there in the 1690s. While the location of the only saltpan to have been built on Skye, by Magnus Prince in 1703, cannot be found, on the more northerly Orkney islands, substantial physical remains of pans first built on Eday in 1633[18] can be added to the cartographic evidence of saltpanning near Kirkwall.[19] Other notable upstanding monuments relating to the industry are to be found on the Clyde islands of Arran and Bute.[20] Other material remains can be seen on Preston Island, on the Forth, which in the early 1800s was probably the only colliery-saltwork in Scotland with a boatman on its payroll; at Joppa and Prestonpans; and further north, at Usan, south of Montrose.[21]

During its 'golden age', salt making was highly visible. Saltworks, enveloped in smoke and steam, were intensely concentrated workplaces at a time when most manufacturing was dispersed in the domestic system. They gave rise to powerful and now forgotten imagery: describing the aftermath of a massive explosion at Kinneil colliery in 1698 to the Earl of Arran, Daniel Hamilton explained that 'the Smoak of Six Salt pans...is not bigger'.[22] The panhouses' visibility made them easy targets for zealous self-styled men of God, who, through the middle decades of the seventeenth century, watched, often impotent in spite of their fury, as the pans continued to smoke on the Sabbath. Especially observant were the elders of Culross and Tulliallan, who not only searched for Sabbath breakers in their own towns but also kept an eye on the south bank of the Forth: in October 1642 they resolved to 'remember the minister to wreat [write] to the ministers of the other syd', as saltpans had been seen 'going ordinarily one the Sabbath-day.' As late as 1677, some members of the Synod of Dunblane were still keeping a surreptitious watch over the pans at Culross and Tulliallan, the ministers of these parishes being asked, 'to learn at the discretest of the salters what is the reason they keip the saltpans ganging so long upon the Lord's Day', and make the Synod's displeasure known to Lord Kincardine, the pans' proprietor.[23]

Rightly, the Forth has been described as the 'chief theatre' of Scottish salt production.[24] For most of the eighteenth century over 90 per cent of Scottish salt was made at pans which lined its shores. It is important to note however that over time marked shifts in the relative importance of the various saltmaking localities occurred. Culross, perhaps the leading salt manufacturing vicinity in the later 1500s, never recovered from the great storm of 1625, which ravaged much of the Forth coal and salt industry. The shortfall in exports was evidently partly made up by the expansion of salt manufacturing further east, at Methil and Wemyss. On the south bank of the river the major exporting contributions came from the pans owned by the Earls of Winton at Cockenzie, in addition to the extensive range owned by the Dukes of Hamilton at Bo'ness. Generally the locational shift was eastwards, down river away from the fresher water of its upper reaches. By the second decade of the eighteenth century, Kincardine and Newpans, to the west of Culross, which had formerly been able to boast some forty pans, were represented instead by seven.[25]

Yet developments elsewhere are also worthy of attention. This is especially true of the string of saltworks which was constructed on the north-east coast, from Dundee to Portsoy, in the years immediately after

1793. Whereas no salt had been made there prior to that date, by the end of the century almost eight per cent of Scotland's salt output came from the north-eastern works. The circumstances of this development, and its consequences, will be looked at in Chapters Two and Four. The latter deals primarily with the relationship between salt and politics. As with the whisky industry, the fortunes of the salt industry can only be fully understood if attention is paid to the legislative framework.[26] A crucial difference though, was that many of the major saltwork owners were also either members of the Scottish legislature, or had influence in it. They even continued to wield enough political clout to protect their salt interests in the British parliament after 1707. From around 1570 until the late eighteenth century the Scottish saltmasters in turn ignored the law and then campaigned to shape it to suit their purposes. Had they not treated Privy Council decrees in such a cavalier manner it is highly unlikely that the industry would have flourished as it did from *c.*1570–1640. Had they not taken steps to change the laws governing the importation and use of salt in the 1660s, it is questionable if saltmaking would have survived as long and impressively as it did. Protected from Cheshire by its geography alone, the Tyneside salt industry, which operated on exactly the same technical principles as its Scottish counterpart, was unmistakably on the road to ruin by the middle of the eighteenth century.[27]

In a perfectly competitive market, Scottish salt would have had few buyers. For the commercial preservation of flesh for export it was virtually useless, and was generally only purchased for this purpose when nothing better was available. The exception may have been cod fishermen, often described as 'north country men' in saltwork account books, who took on board salt from pans at the seaward end of the Forth, such as Cockenzie and Dysart, on their way to the fishing grounds around Iceland. In the early 1770s, during a particularly buoyant period for the Peterhead cod fishing industry, cargoes of several hundred bushels were frequently dispatched northwards from Bo'ness, and in 1787 it was reported that fishermen were taking 'much salt' from Cockenzie and St Monance saltworks. For preserving oily fish such as herring and salmon however, which had to be more heavily salted and saturated in brine, Scottish salt was rarely used. Salmon curers such as John Richardson of Perth, 'the Great Fishmonger of the Tay', only used imported salt, mainly from Portugal. The Shetland fishing industry seems to have been driven to use substantial quantities of Scottish salt only once, in 1780–1, during the Dutch War.[28] For those with no choice though, such as seamen, it was often purchased, to preserve ship's stores of beef and pork. Sea salt was unsatisfactory in several ways. Not only was it liquescent, but it contained several impurities which made it less desirable than other salts. Sir George MacKenzie of Rosehaugh for instance, in his *Memoirs* of the late seventeenth century, complained that, 'if hides be cured with our salt, they...fall out in holes'.

Contemporary chemical analyses merely confirmed what its more critical users suspected. 'Scotch Common salt' in the early nineteenth century was found to contain almost five times as many impurities as did 'Cheshire

Fishery Salt', and a considerably greater volume of these than were found in 'Bay' salt. The impurities referred to were 'uncombined magnesia', which gave the salt a bitter taste, sulphate of lime, sand and what was described as 'coaly matter'.[29] A problem for all saltmakers, observers such as William Brownrigg thought that the particularly objectionable 'dirty grey colour' of Scottish salt was largely the result of the salters' failure to 'clarify the brine'.[30] This was a means used early in the boiling process to remove the worst impurities by adding coagulants such as ox's blood or eggs to the heated water and then drawing them off. How far Brownrigg's 1748 analysis was based on personal observation,and how much he was influenced by the virulent anti-Scottish sentiment of his age is not clear: blood and eggs were certainly used at a number of Scottish saltworks.[31] What does seem likely, however, is that generally speaking, less concern was paid to the quality of Scottish salt after the middle of the seventeenth century. As will be shown in Chapters Two and Four, the saltmasters then began to secure a monopoly of the home market. Paid by the quantity of salt they produced, as long as it was usable and had a sale, salt workers were little concerned about its quality. They were given little encouragement to evaporate the water more slowly, which would have cleared it better and produced the bigger grains of salt which were sought by commercial preservers of meat and fish.

Contemporaries then, looked enviously at the methods used to make salt elsewhere. Instead of the single day it took to make a pan of salt in Scotland, the Dutch took six, and made salt which was eminently suitable for preserving fish.[32] French salt, reported Sir John Lauder in the 1660s, 'is a great deal better and more savoury than ours is.' Like many others he was struck by the comparatively low production costs of the 'Bay' salt producers, so named because of the early concentration of salt making in this way on Bourgneuf Bay, south of the river Loire, in France. Unlike the Scots, saltmakers on the Atlantic coasts of France, Portugal and Spain used no coal to make salt, and relied instead on the favourable operation of the elements. 'In 2 heurs tyme the sun will converte a great ditch full of sea water into upright salt', Lauder wrote. In '3 moneth, May, Juin, July, they [the French] make more salt than the fire maks in 2 years in Scotland'.[33]

Also cheaper to convert to usable salt were the brine springs and rock salt deposits of Cheshire. Here the ratio of coal to salt was 1:1; in Scotland it could be as high as 8:1. It was little wonder then in the later eighteenth century, as duties on legal sales of salt rose, that an illicit trade in rock salt developed. The latter was shipped legally and free of duty from Liverpool to Ireland, where it was refined, and then smuggled over the Irish Sea into the west coast of Scotland. If the 9th Earl of Dundonald's estimates of the size of this trade are to be believed, the Scots illegally imported more salt via this route than they made.[34] Even though his figure of 460,000 bushels per annum of smuggled salt was probably exaggerated by his anxiety to convince the authorities of the baneful effects of the salt laws,[35] it does serve to emphasise the relative inefficiency of Scottish marine salt manufacture. In the 1780s a bushel of 'small' salt in Liverpool cost 8*d.* On the Forth, the same quantity would have cost, at the pans, 1*s.* 3*d.*[36] Notably, most of those

Scottish saltworks which survived the repeal of the salt duties in 1823, did so in part by using imported rock salt and refining it in salt water. What few attempts were made to improve the quality of Scottish salt, and the reasons for their failure, are discussed in Chapters One and Four.

No study of the Scottish salt industry would be complete without paying considerable attention to the salt workers. Indeed without some effort to describe and understand something of their role and experience such a book would be without the human interest which lies at the heart of all but the least consequential history. The salt workers have usually been treated as collier-clones. In one important respect, this practice can be justified. From 1606 both colliers and salters were treated as a form of serf, legally bound to their places of work for life. Later, they found themselves subject to parliamentary regulation of their conduct at work. How much this catalogue of intended oppression affected the salters is examined in Chapter Five. Often casually described as a 'race apart', some consideration is also given to the question of their occupational and social status.

At this point however, it is worth commenting briefly on the aforementioned tendency to lump the work experiences of the colliers and salters together. It has to be admitted that it was not only their servitude they shared. Invariably in the coastal regions where a colliery and saltwork were linked, the two sets of workers lived in the same locality. This did not necessarily mean that they shared a separate existence though, or felt a greater affinity towards each other than did, say, colliers and weavers. The colliers who worked Wemyss of Bogie's coal in the early eighteenth century, for instance, did not live with the estate's salters, but in the town of Kirkcaldy, scattered amongst the urban population. At Culross, colliers and salters were housed in separate rows.[37] The offspring of the families of colliers and salters did frequently intermarry, but as recent work on the marital patterns of coal miners has shown, by no means did they restrict themselves to partners from the same occupational grouping.[38] Firmer comments on the extent of intermarriage between the sons and daughters of colliers and salters await further systematic investigation of parish registers.

Their very different workplaces and work routines also acted to keep the colliers and salters apart, and to provide each of them with a distinctive work experience. The former worked underground, on a daily basis, coming up from their pits in the afternoon. Salters worked above ground, usually in groups of two or three, in panhouses which could be two or three miles from the pits. Saltmaking was a continuous process, which kept the salters at the pans for twenty four hours or longer at a single stretch. Work, sleep and leisure were less easily compartmentalised. As far as the working periods of the lives of colliers and salters were concerned, there was little or no direct communication between the two groups. Indirectly however, tensions could arise. Concerned to put out as much coal as they could, in April 1719 the Bogie colliers were sending up pancoal mixed with 'poor' coal. So angered were the salters, whose output of salt was partly dependent upon the quality of coal they received, that the estate's tutors were forced to ask their factor to hear both sides in 'all maters of this kind betwixt Coaliers and Salters',[39] suggesting that disputes between the two groups may not have been infrequent. That colliers and salters had much in common should not be allowed to conceal the differences that existed between them, or the points at which collier and salter society merged with that of the rest of the labouring poor in Lowland Scotland. This theme will be further explored in Chapter Five.

This book then attempts to cover a great deal of ground. It is not claimed to be definitive: as the first book-length study of the Scottish salt industry, which attempts to deal with a period of three hundred years, this would be foolishly optimistic. Hopefully though, the principle objectives which have guided its preparation have been met. First, that a relatively clear picture of the course the industry followed over time has been described and accounted for. Hitherto this has not been clear,and considerable confusion has arisen over the industry's fortunes from the middle decades of the seventeenth century. Second, and equally important, it is to be hoped that despite the necessary division of the book into chapters, the various components of the industry with which these deal will be seen as an interlocking whole. Labour, for example, is often treated in industrial histories written by economic historians as a 'dependent variable', of only passing interest compared to the mighty concerns of capital sources and entrepreneurial skills. Nor is it unknown for an industry's trading fortunes, its markets, to be accorded a secondary role. The history of salt manufacturing in Scotland, like other industries, cannot be fully understood without an appreciation of the diverse forces which influenced it: ranging from European wars and their impact on trade routes, through domestic politics, to local weather patterns and the skills, cultural values and attitudes to work of the salter serfs. The book has been written by an academic historian in an academic institution. It is to be hoped that fellow scholars will judge that it has been prepared with sufficient rigour. Its purpose will not have been achieved however if it cannot also be read with pleasure by a wider audience. Its third aim has been to attempt to recreate as vividly as possible with the printed word and a few illustrations, some of the principal elements of a once vital but easily overlooked aspect of Scotland's past. If this has been anywhere near satisfactorily achieved, the book should be of interest to anyone with a serious interest in Scottish history.

1

MAKING SALT: WORK PROCESSES AND PROBLEMS

GENERAL SURVEY

The manufacture of salt, by boiling sea water, was always a rather primitive technological process. At most works, sea water was collected in rock-cut or stone-built basins or ponds which were widely known as 'bucket pots', although the use of hand buckets to remove water into the pans, the process which gave them their name, became increasingly rare. The bucket pots, filled with water which ran through channels or water courses at the flood tide, were used both as a form of settling tank, wherein sand, mud and other extraneous matter could fall to the bottom and thus be kept out of the pan, and as reservoirs to provide water supplies for the pan or pans when the tide was out. Some slight evaporation also took place. From the bucket pots the water was transferred, either by hand or wind and later, occasionally, by steam pump, through what were often called 'pan spouts', that is wooden, and later iron, pipes, into a wooden, iron or stone cistern, or 'pond' which sat near to the panhouse, prior to being drawn into the pan itself. Although some uncertainty surrounds the matter of their size, from the early seventeenth century most pans appear to have been similar to those described by the Swedish industrial spy Henry Kalmeter in 1720:

> The pan itself is 18 foot long and 9 foot wide, and is made of iron plates one ell or more long and one quarter or more broad and half an inch thick: these are fastened together with strong nails, and supported with bars to prevent their weight buckling them. The pan is 18 inches deep, and is filled with water to 15 inches.[1]

Panhouses were usually stone-built structures, with steeply-pitched turf or heather-covered roofs, although for the latter purpose tiles became common from the later eighteenth century. In the pan the water was heated, although prior to boiling, quantities of egg whites or bullock's blood were usually thrown in, and

> ...by a sudden transformation the seething white liquor became increasingly overspread with a thick variegated or brown bubbling scum. The albumen in the blood, which had coagulated and risen to the surface carrying all the impurities with it, was then skimmed off, leaving the boiling brine clear as crystal and devoid of suspended matter.[2]

This done, evaporation proper could begin, with 'panwood' or small or dross coal generally being the fuel. The ratio of coal to manufactured salt was around 6:1, occasionally 8:1, but never as high as the 16:1 mentioned by some authors. This however was still six or eight times as much as was used by the Cheshire brine spring and rock salt refiners.[3] The evaporation process took between 22 and 28 hours, during which time the pans could be refilled three, four or even five times (depending upon the salinity of the sea water) until a 'full' or whole pan of salt had been made. The hot salt was then drawn to the sides of the pan, allowed to drain for a short time and then removed in baskets, 'pocks' (small sacks) or boxes to the girnel (storehouse).

A brief description such as this however, although essential at this stage, is wanting, not least because it cannot incorporate the wide variety of local and regional circumstances and techniques. Of necessity it overlooks most of the modest but nevertheless perceptible changes in production methods which occurred between the sixteenth and nineteenth centuries. Above all perhaps it conveys the impression that salt manufacturing was rather straightforward, and by so doing conceals the many problems faced by the salters, the elucidation of which is essential if a fuller understanding of the work process and output patterns of the industry is to be achieved.

From the fifteenth to the nineteenth centuries, the vast bulk of Scottish salt was made in coal-fired iron pans, most of which were situated on or near the coal-bearing land which stretched along both banks of the river Forth.[4] (See Maps 1 and 2) In the early eighteenth century for example well over 90 per cent of salt made in Scotland came from this region.[5] Coal-fired salt pans were also to be found in Ayrshire, Kintyre, on the island of Arran and at Brora in Sutherland. Yet peat and turf were also used to fire pans, mainly in Galloway and on Orkney, which it must be stressed, accounted for a very minor share of recorded Scottish output, probably never more than one per cent between them. Galloway's actual share however may have been marginally greater than is indicated by this figure, owing to the fact that the salt made in Annandale, to the west of the region, was relieved of the burden of paying salt duties from 1671[6] and consequently any salt made in this district was not recorded in the official totals.

As in the coal-using salt manufacturing areas, the longer term success of salt making in Galloway and Orkney was heavily dependent upon the ease with which ample supplies of cheap fuel could be obtained. It seems more than mere coincidence that those works which survived longest were situated near to deep and extensive peat deposits. This was true of the Annandale parishes of Cummertrees and Ruthwell for instance, where salt making lasted at least until the second decade of the nineteenth century, and at Galdenoch, to the west of Stranraer, where salt was still being made in the 1790s, some decades after most of the other works in the area had disappeared.[7] Similarly, salt making in Orkney was carried on longest on Eday, which was a supplier of peat not only to other islands nearby but also, in the 1840s, to Edinburgh.[8] One of a group of twelve pans which had been constructed at Calfsound of Eday in 1633 was still producing salt in the second half of the eighteenth century. Saltpans were also built on the peat-rich islands of Cava and Flotta, as well as at similarly well-endowed Orphir, near Kirkwall.[9]

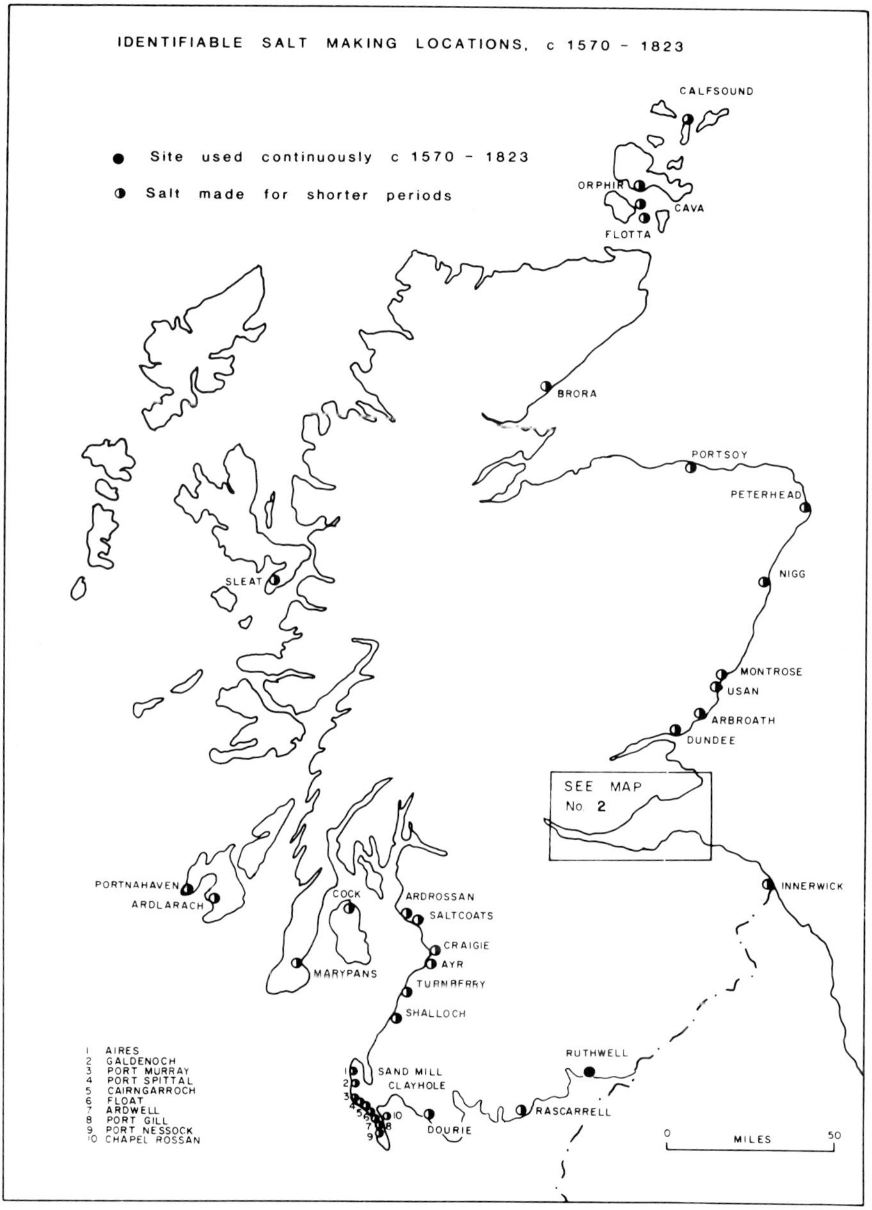

Map 1 Scottish salt manufacturing locations, *c.* 1570–1823.

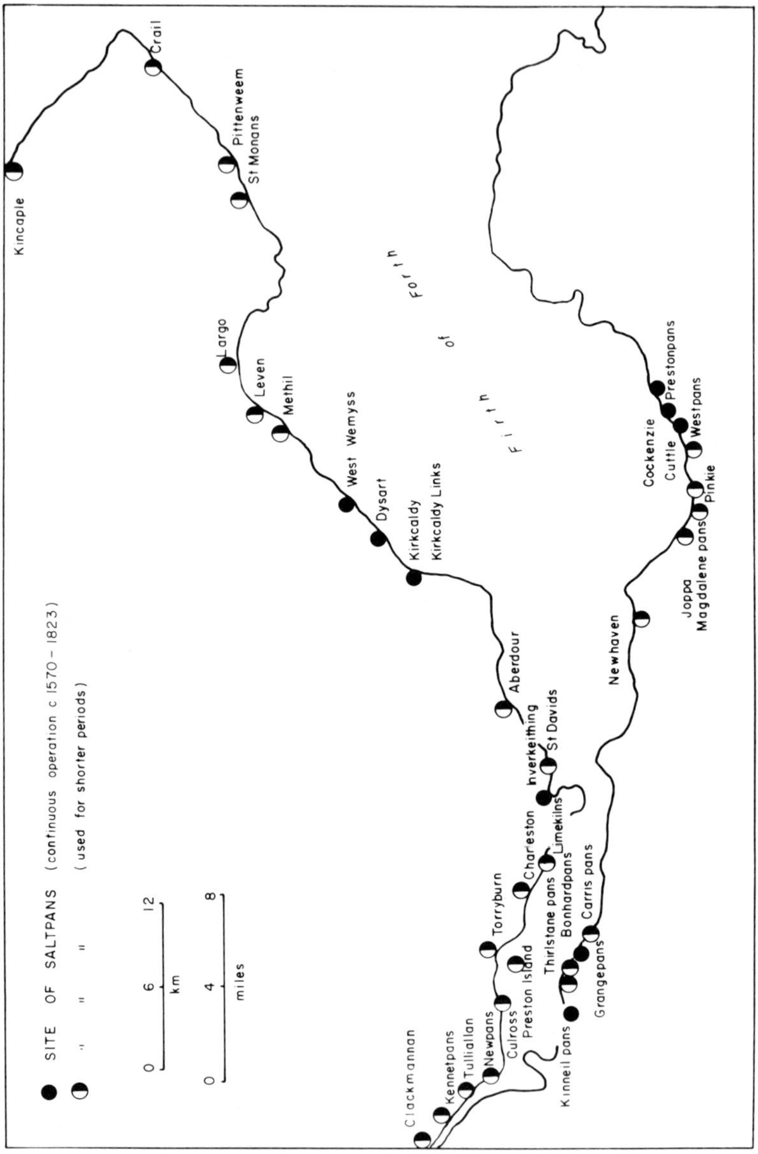

Map 2 Salt manufacturing locations on the Forth and its estuary.

In sharp contrast to the coal-fired industry however, where turf or peat was used, salt making was normally very much a part-time seasonal operation, requiring significantly less capital investment. Unlike the salters of the Forth and Ayrshire, who depended on saltwork employment for the greatest part of their incomes, salt making in Galloway and Orkney was always a subsidiary occupation. There was less distinction between pan owners or lessees and operatives. Thus it was usual in Galloway for tenant farmers to rent a saltwork as part of a lease of a farm or other enterprise, as in the case of Patrick McNeilly, who in 1741 leased the lands and corn mill of Meikle Galdenoch along with 'liberty to Build and keep up a Salt pan'.[10] A later tenant of the same lands combined agricultural work and salt panning with small scale fishing.[11] Cottars provided the occasional labour required at the pans. As a result the experience of salt making here was markedly different from those areas where the industry was carried on full-time. In Cummertrees in the early nineteenth century for example, the salt makers were drawn from the ranks of tenants of very small acreages of arable land, the working and improvement of which they combined with peat digging and its sale in Annan along with short periods of salt making. A wet summer might mean they made no salt at all, while in 1714 it was reported that the manufacture of salt in the district rarely went on for more than four months.[12] The average time spent on the task, it was claimed in 1812, was between twelve and fourteen days.[13] The lowly social status of the twenty or so 'poor' families who were thus engaged in keeping their heads above subsistence level was evidently not shared in neighbouring Ruthwell parish however, where salt making was apparently in the hands of more substantial farmers.[14]

Technically, making salt with the use of turf or peat fuel was very similar to the process outlined at the beginning of this chapter. Only in Annandale was the procedure notably different, with the salt being distilled from 'sleech' or salt-impregnated sand which was gathered in dry weather during the summer months and taken on horse-drawn sledges to the pans, which until the early 1800s were made of lead.[15] Being only '4 foot long, 3 foot broad and 4 inches deep' from the seventeenth until the nineteenth century, and in spite of the fact that they were always used in pairs, the scale of operations was substantially smaller than on the Forth, where pans were generally four or five times larger.[16] Whereas an average sized saltpan on the Forth in the eighteenth century could, if in good condition and kept going continuously, produce in the region of 4,500 bushels per annum[17], or at least 60 bushels per week, the smaller Annandale pans could at best hope to put out 20 bushels in a week.[18] Actual output was often far below this.

Extremely low output figures were also the norm elsewhere in the south-west, and are a further indication of the shortness of the salt making season. On average, the seven saltworks which were listed as 'going' in Dumfries and Galloway in 1716–7, sold 57 bushels of salt per annum: in the same period annual sales from the country's largest saltworks, at Cockenzie on the Forth, were in the region of 20,000 bushels.[19] The capital invested in plant and equipment was proportionately less, with panhouses and ancillary structures usually being smaller and less substantial than in those areas where coal fuel was used. Perhaps representative of the situation in the south-west is a valuation of Galdenoch saltworks in 1741:

> The two dwelling houses at the Salt pans are valued at Two pound Sterling. The pan house & peat house and a larger Hole digged in the rock...ten shillings.[20]

1 An eighteenth century depiction of the salt manufacturing process. The pan was suspended by hooks from a wooden panstand. The underpan furnace can also be clearly seen. The salter is seen removing salt from the pan. The relatively light and spacious working area suggests that the picture may be rather idealised or of a construction not commonly found in Scotland. Stone-built and turf- thatch- or slate-roofed Scottish panhouses of the period were hot, steamy, dark and cramped places in which to work. From W. Brownrigg, *The Art of Making Common Salt* (1748).

In the later seventeenth century and for much of the eighteenth, the capital cost of a single pan and panhouse elsewhere in Scotland was reckoned to be in the region of £150.[21] One partial consequence of this in Galloway was that while a fairly large number of saltworks was established, especially from the later 1600s until the mid-eighteenth century, few remained operational for long, their owners apparently being able to abandon their relatively slight investments when either the circumstances of peat supply or market demand turned against them.

Not all peat-fired saltmaking ventures were on this modest scale however. Inspired by the appearance of market conditions which were highly favourable for Scottish salt producers (see Chapter Two), between 1692 and 1694 Alexander Campbell of Cawdor lent James Scott, a merchant and already the 'Master of ye Saltpan far Islay' at Portnahaven, 3,000 merks for the construction of three pans and 'other necessare buildings' at Ardlarach.[22] In return for this, and 'the liberty of moss in the said Towne' Scott was to pay an annual rental for each pan of 200 merks with three bolls of salt for 19 years, during which period he was to be responsible for the upkeep of the works and ensure that a supply of 'Servants and workmen' was maintained.[23] Scott however appears not to have kept his part of the bargain and the scheme failed, although salt continued to be made on the island until the 1720s.[24] Further north, on Skye, Sir Donald MacDonald of Sleat also demonstrated some interest in large-scale salt manufacture and in 1703 advanced £1,400 Scots to Magnus Prince to erect a saltpan.[25] Unfortunately, little more is known about this scheme. Industrial archaeological evidence from Eday though, where substantial traces of the two panhouses used in the eighteenth century still survive, indicates that the iron pans there would have been of a comparable size to those used in central Scotland in the same period.[26]

Most Scottish salt however was made in coal-fired pans, and it is to these that most attention must be directed. While work went on round the year, as in much early industry in Britain, production was far from regular and was subject to a number of interruptions. As far as the salt industry was concerned, the Act passed by the Scottish Parliament in 1641 which insisted that colliers and salters should work six days of the week, on pain of a fine or 'other punishment of their bodies', was virtually a dead letter. While many if not all of Scotland's saltmasters may have wished to induce more regular working habits in the pans, inherent in the salt making process were too many factors over which they could exercise little or no control.

From wherever adequately detailed output records have survived, the same picture emerges, of broken production runs and long periods when pans were temporarily silent. At Bogie saltworks near Kirkcaldy in the twelve months from October 1720 for example, in only 25 of the 52 weeks were all five pans going at once.[27] This was by no means unusual. In the year from October 1724 average output per pan per week was 2.27 'draughts', or full pans of salt: optimum output was four draughts. Only 56.76 per cent of the works' potential salt production was achieved. Only in 26, or 10.20 per cent, of the possible 'pan weeks' did the salters manage to make four draughts, while in a slightly higher proportion of weeks, 10.98 per cent, no salt at all was made at one or more of the pans. In spite of claims that the pans at Winton saltworks at Cockenzie on the south bank of the Forth were better managed,[28] production was still subject to lengthy stoppages, as Table 1.1 reveals.

Table 1.1

Winton Saltworks: Number of Weeks Worked by Salters, 1716–8 (July–June)

Salter names	1716–7	% (of 52)	1717–8	% (of 52)
John Greg	21	40.38	35	67.31
Archibald Greg	26	50.00	37	71.15
William Greg	34	65.38	28	53.85
Robert Donaldson	31	59.62	33	63.46
Alex. Mathieson	55	67.31	29	55.77
Christopher Fleucher	39	75.00	23	44.23
David Balvaird	30	57.69	30	57.69
John Donaldson	13	25.00	25	48.08
George Hunter	28	53.85	30	57.69
Nicholas Watson	36	69.23	29	55.77
James Brown	37	71.15	26	50.00
James Watson	22	42.31	25	48.08
Average	29.33	56.41	29.17	56.09

Source: SRO, FEP, 1715/E661/Box 196, Winton (Cockenzie) Saltwork Output Books, July 1716–June 1718

The average number of weeks worked by the salters in each of the two years for which data is available was just over 29, or 56 per cent of the annual 52. The various salters' experience was by no means uniform however; while in 1716–7 James Brown managed to put out salt in 36 weeks, John Donaldson was only able to work his pan for 13.

While strenuous efforts were made by saltmasters and their grieves to reduce the number of interruptions, results were only achieved slowly. Thus at Dysart saltworks in 1817, during the most favourable period of the year, from May to November, the proportion of panweeks in which no salt was made was 7.05 per cent, and while each salter worked for at least some part of most weeks, full output was achieved in only 52.56 per cent of them.[29] Production then was still subject to frequent stoppages, but for shorter periods, of hours and days rather than weeks and even months. Paradoxically, what the limited surviving evidence concerning work patterns later in the nineteenth century suggests, is that it was in the industry's twilight years, when regulatory legislation passed by the old Scottish parliament no longer applied, but when commercial pressures were acute, that salt panning was more likely to have been carried on for six days of the week.[30] Such a conclusion should be treated cautiously however; circumstances could vary between both individual salters and saltworks, and a period of intense demand, as in the 1630s and 1640s, could cause the pans to be worked continuously, even, it seems, on the Sabbath.[31]

Intensive periods of salt making however could not last for long, owing principally to the constraints imposed by technology, nature and the seasons, and the social values and customs of the salt workers. Statistically the least significant, the latter category will be looked at in passing only here, and examined in more detail in Chapter Five. Repairs, to pans, panhouses and bucket pots, accounted for around half of the stoppages in production at Scottish saltworks. As Table 1.2 reveals, at Bogie this was by far the most important reason why the salters failed to achieve at least three draughts of salt weekly in the year from October 1724.

Table 1.2

*Bogie Saltworks: Reasons for Non-achievement of 3/4 Draughts of Salt per Week, 1724–5 (October–October)**

Category	Number	%
Pan repairs (including storm damage)	48	57.14
Water shortages	6	7.14
Salter shortages**	7	8.33
Lack of coal	5	5.95
Sickness	4	4.76
Family commitments	2	2.39
Unknown or unclear	12	14.29
	84	100.00

*reasons only given where less than 3 draughts achieved
**generally due to absence of salter assistants

Source: NLS, Wemyss of Bogie MSS, MS 3088, Bogie Salt Book, 1720–32

The picture was very similar elsewhere. At Winton for example pan repairs accounted for 45 per cent of the weeks when little or no salt was made in 1716–7.[32] Nor was the problem easily overcome; at Dysart in the early nineteenth century, the necessity of repairing pans accounted for over half of the occasions when there were significant shortfalls in output, as the following table shows.

Table 1.3

Dysart Saltworks: Reasons (Where Given) for Significant Losses of Weekly Output, May–November 1817

Category	Number	%
Pan repairs (including storm damage)	40	54.05
Lack of coal	23	31.08
Sickness	5	6.76
Family/community obligations	4	5.41
Miscellaneous/unknown	2	2.70
	74	100.00

Source: Wemys Castle MSS, Oncost Book, Dysart Saltworks, 3 May 1817–7 February 1818

Some repairs or pan maintenance had to be done regularly and were entirely predictable punctuations in the salter's calendar. At what were usually monthly intervals, each pan was stopped and cooled down while 'paddling' or 'paidling' took place, an operation which lasted one or two days. Artificial evaporation of the sea water caused the sides and bottom of the pans to become 'crusted over with some sort of matter' (calcium sulphate) which, if not removed, burned and eventually holed the pan plates, necessitating their replacement, at the proprietor's expense. It was apparently to avoid this and encourage his salters to remove what was sometimes called 'stone scratch' with picks regularly, that David, 2nd Earl of Wemyss, in 1659, arranged that they should each be given a quantity of free coals and 24*s.* Scots every month, 'in respect...they have charges in paddling the pan and putting pitch on her.'[33]

'Beiting' was the other regular interruption to the salt making process, a task made necessary by the scaling and burning which corroded the pans and panstands over long periods of constant use. The pans thus had to be stripped down and repaired, with plates and nails being replaced as required, and then rebuilt. At most works beiting appears to have been done at least every two years, generally of necessity, but also in part owing to the force of custom, the salters at Cockenzie in the 1720s for instance insisting on the break even though the method of firing there apparently did less damage than was the case in east Fife, where the pans 'when going to beet are all miserable Scorch'd...or burnt'.[34] The time a pan was stopped 'at beit' depended upon a number of factors, such as the extent of the damage, the availability of materials, notably iron plates and nails, and whether suitable smiths and other tradesmen could be found.

The job could rarely be completed in less than three weeks, and often it took longer, six or eight weeks, although stoppages of this duration became

2 Upstanding remains of panhouse, Ascog, Bute. The water inlets can be seen on the rounded seaward gable wall. The building, which may never have been used, owing to a local shortage of coal, appears to have been built in the mid-seventeenth century. The rounded gable end and its internal dimensions are typical of early Scottish panhouses, similar traces of which can be seen on the islands of Arran and Eday, Orkney. The Ascog remains are the most complete for the period. Photograph by John R. Hume.

less common after the middle decades of the eighteenth century as saltpan proprietors became increasingly cost- and efficiency-conscious. To hire individually the six or seven smiths required to beit a pan had long been recognised as wasteful, as they had the 'opportunity to be slow in Working As Also of Abstracting the Iron both old and New', and from the early eighteenth century and occasionally beforehand the practice of sub-contracting the work to a 'principal smith' and his assistants had been adopted.[35] The main drawback of this system however was that the payment of the individual daily wages of the principal smith and his assistants (£1 and 15*s*. Scots respectively at Winton in the 1710s) was still the responsibility of the saltwork proprietor or grieve. In consequence saltwork employers sought to introduce a less costly arrangement, such as that entered into between the proprietors of Grange colliery and saltworks and John Patrick in 1781, whereby the latter, a smith, contracted to beit the five pans for 50*s*. a month, and pay his assistants himself. Efficiency was to be further improved by the provision, for Patrick's use, of 'a machine for turning up the pan to beet',[36] which hitherto had been a cumbersome and time-consuming operation.

Rather less easy to forecast was the appearance of serious cracks and other flaws in the panhouse masonry. Usually the result of the continuous heat which was required to make salt, furnaces, hearths and 'lums' all had to be periodically repaired, although the problem generally had considerably less impact on output than did beiting. Stoppages for this type of routine masonry tended to be for days rather than weeks, although major building work could take longer. One of the most common building jobs was to replace the panhouse hearth, for which at Winton in the 1710s a more or less standard fee of £11 Scots was paid, compared to the sums in the region of £100 Scots which the smith work alone (excluding materials) cost for a pan beiting.[37] Stone work was usually done by masons hired on a daily basis, although this was not the practice at Kirkcaldy, where the salters built their own hearths. Internal timbers too quickly corroded: as a nineteenth century architect commented after inspecting the pans at Pinkie: 'From the state of temperature arising from them [the pans], it is not expected that the Timber is to remain long in a healthy state'.[38]

A greater threat to panhouse structures came from storms, which periodically wreaked havoc on buildings which of necessity were sited at exposed coastal locations. Perhaps the best-known was that of 1625 which was said to have destroyed most of the saltpans (and some collieries) on the Forth, including the then important group at Culross.[39] Another, in January 1752, caused severe structural damage at Dysart, where a small army of quarriers and sixteen masons had to be hired to 'make up the breatches thatt the Storme Meade upon the Sallt pans'.[40] Roofs were frequently torn off, with the result that production was often stopped for several weeks, as during October and November 1716, when David Balvaird, a Winton salter, was unable to make any salt, although he was more fortunate than his fellow salter Archibald Greg, whose pan roof was blown off later in October and not replaced until the following March.[41]

3 Rear view of remains of panhouse at Cork, Arran, constructed *c.* 1710. The building is divided into two sections. The saltpan itself was located in the room nearest to the sea. The nearer working area was where the underpan fire was stoked from, and in which coal was stored. Note how steeply pitched the heather and turf roof would have been, which contributed to the small internal working space. The fine curved gable wall has recently collapsed. Photograph by C. A. Whatley.

While the reasons for this unusually long stoppage are unknown, the length of the more normal shorter periods, of anything up to six weeks, can be accounted for by the time taken to remove the old roof timbers, arrange the purchase and preparation of suitable new wood, 'double trees', 'single trees' (beams) and 'dails' (planks), and then employ wrights to put the new roof on. The pace of work on the roof itself was largely determined by the latter, to whose daily rates of pay of 12*s*. Scots in the early eighteenth century have to be added the customary charge on the saltpan proprietor of 16*s*. Scots for a gallon of 'pan roof ale', when the job was completed.[42] The cost of the timber and some pieces of iron was not insignificant; over £169 Scots was paid for these when David Balvaird's roof was being replaced.

Storms, wind and rain also adversely affected the salt making localities in other less obvious ways. During the late autumn and winter months salters were pestered by seaweed and other unwanted material being driven into the bucket pots, rendering the water unsuitable for salt making. In the twelve months from July 1716 all but two of the thirteen pans at Winton were stopped for this reason, some more than once, and in the case of James Watson, three times, the last after an early summer storm in June 1717.[43] This was not unusual, nor does the problem appear to have become any less severe over time; at Dysart a century later for instance groups of four salters were frequently sent out to clear out or 'cast' one of the bucket pots, in return for a payment of one shilling each.[44] Rain too could halt production. On the frequently rain-drenched island of Arran, the saltwork belonging to the Dukes of Hamilton was plagued with rain water which not only periodically flooded the coal pits which supplied it, but also streamed through the panhouse roof, which in time forced the salters to temporarily cease work.[45] The problem was that the saline content of the sea water in the pan was diluted, making it difficult to make good salt, and adding substantially to evaporation costs as it took longer to produce a panful of salt. On the east coast rain water threatened production from a different direction, by what Thomas Tucker in 1656 had identified as the 'suddaine coming downe of the freshes'[46], where heavy rainfall swelled streams and rivers in the locality of saltworks and consequently caused the difficulties already described.

Partly for these reasons, it was usual for salters to be allowed an additional quantity of coal for their pans during the winter months. At Bogie for example from November 1730 until the following February the salters were given eight instead of the usual seven 'dozens' of coal for which they were obliged to return three bolls of salt.[47] Similarly at Grange saltworks near Bo'ness it was agreed with the salters in 1801 that during the winter months, 'in time of great rains', they should have eleven rather than ten carts of coal weekly.[48] 'Freshes' continued to exert their unwanted influence through to the middle of the nineteenth century and beyond. Thus as late as 1845 the salt agent at St Davids was anxious to obtain an additional shipment of rock salt from Liverpool to mix with the local brine, owing to the 'great quantity of Snow' which had fallen and weakened the water of the Forth.[49]

Lack of water could be equally distressing for the saltpan proprietor. As Table 1.2 revealed, some seven per cent of the serious shortfalls in output from Bogie in the mid-1720s were due to this, 'because the neep was so very low that it could not reach the potts'.[50] The problem however had largely been overcome by 1732 as the affected bucket pots were rebuilt and enlarged, 'especially towards the Sea, in order to give them brue in time of Neeps'.[51] Less easy to resolve satisfactorily was a dispute involving David Tilloch, in the West Pan at Bogie. His bucket pot was within the town lands of Kirkcaldy, and a number of local fishermen in 1725 were alleged to have been using the pot as a store for their mussel bait. In consequence there was insufficient water, 'for the Service of the Pan betwixt floods or Spring Tyde', while the available water was so 'foul'd & puddled with the slime & trash' that Tilloch refused to go on making salt.[52] While the latter difficulty was probably unique to Bogie, most works at one time or another were temporarily crippled by leakage from the bucket pot, either through a disintegrating or collapsed retaining wall or because of the failure of the wood and leather sluice gate mechanisms which allowed water to flow in to the reservoirs when the tide flowed, and to prevent its departure during the ebb.

All works were subject to the threat of either temporary or permanent closure due to the absence of coal. Over the course of the three centuries with which this book is concerned, salt panning disappeared from a number of locations as coastal seams were exhausted or local pits flooded. The economics of the industry were such, that with the exception of extremely buoyant periods of demand, coal fuel had to be available at minimal cost, which meant that its transportation from pits which were much more than a couple of miles from the pans was deemed impracticable. Thus insufficient drainage and the subsequent failure of Sir Peter Wedderburn of Gosford's coal at Innerwick in the 1680s (see Map 1) meant that there was no alternative but to abandon the two pans.[53] Similarly the Duke of Argyle's saltwork at Marypans near Campbeltown on Kintyre had to be given up some time in the later 1750s after a final unsuccessful attempt was made by William and Samuel Mitchell to work the poor, sulphurous and wet coal which had served the pans.

Perhaps the most spectacular reduction in salt making activity occurred at Culross after the devastating storm of 1625. Prior to this it was claimed that Sir George Bruce's salt panning empire amounted to some forty-four pans, putting out 90 tons of salt each week between them.[54] The major colliery there seems not to have recovered fully after 1625 and in the century from *c.*1676 was apparently given up. Although coal and salt operations were restarted in 1774, both were in a poor condition by 1795, when it was reported that the colliery steam engine had lain idle for over four years and that the five salt pans were in various states of disrepair.[55] Further west, along the river bank at Newpans, only the walls of the four panhouses remained.[56] Only in a few cases where coal supplies in the immediate vicinity of the pans had been exhausted was salt making continued. Almost invariably this was when the location of a saltwork coincided with the route of a waggonway where the primary purpose was to carry inland coal to shipping ports, as for example between Tranent-Cockenzie (1722,) Fordell-Inverkeithing (*c.* 1770), Halbeath-St Davids (1780–3), and Stevenston-Saltcoats (1812).[57]

Rather less dramatic were the temporary interruptions in coal supply to the pans. While the incidence of this problem varied both between places and over time, its impact could be substantial and was often the second most likely cause of stoppages to salt production. As Table 1.3 showed, at Dysart in 1817, 31 per cent of significant losses of output were due to coal shortages, which was also the reason for almost 20 per cent of the 'silent' weeks at Winton saltworks a century earlier. Certainly up until the end of the eighteenth century seasonal factors were clearly at work. Thus of the 54 panweeks at Winton which were lost between 1716–7, over half, 56 per cent, were the result of the salters' 'leaders', or carriers of coal from the pits to the pans, going off for a week or two in May and September, to attend to their small portions of agricultural land at seed and harvest time.[58] Periods of heavy rain too could make land transportation of pancoal, even over short distances, difficult if not impossible. Even prior to their closure Marypans saltworks on Kintyre had never been particularly successful, being 'frequently idle' during the winter months, not only because of the quantity of fresh water in the area but also because while the pans were only a mile from the pits bad weather hindered the carriage of coals.[59] Winter rains too frequently overcame the efforts of hand windlassmen and horse driven gins to keep mining operations going, and consequently salt making was brought to a halt.

Less significant but an all too easily overlooked reason for the temporary involuntary cessation of salt production was the incidence of sickness amongst the salters. Although most absences for this reason were brief, lasting a day and certainly less than a week, more substantial periods of sickness did occur at virtually every works for which adequate records have survived (see Tables 1.2 and 1.3 for example). Unfortunately little is known, nor can be told from currently available sources, about the nature and causes of illness amongst this group of workers. With the rest of the population they obviously suffered from the periodic visitations of the myriad of diseases and ailments which, if not fatal, could in early modern Scotland, cause almost unimaginable pain, discomfort and debility.[60] Rarely however do saltwork account books record a salter's absence for this reason by anything more illustrative than the brief description, 'sick', or 'unwell'. How far salters were afflicted by occupationally related disease or injuries is uncertain, although evidence from Cheshire in the nineteenth century, where the evaporation process was very similar to that used in Scotland, suggests that saltworkers were particularly at risk of falling into the boiling brine due to their cramped working space and the fact that the pan rim rarely rose above knee-height, thus making it difficult for them to keep their balance.[61] This may well have been the cause of an accident at Bogie pans in 1730 when it was claimed on behalf of the salter James Rae, who had 'no hopes of his recovery though he has imployed the best

Surgeons', that his injuries had resulted from 'a fall from one of the pan Stands whereby he has been disabled these ten weeks'.[62] That work-related accidents were not infrequent is also suggested by the payment by the proprietors of Dysart saltworks in the early nineteenth century of a quarterly fee of six shillings to a Dr Law, 'for Medicine & attendance to Salters'.[63] Pulmonary problems are likely to have been present too: one visitor to Scotland in 1784 was struck by the corroding 'vapour' of the pans, which was 'somewat injurious to the lungs'.[64]

Thus far this account has been primarily concerned with factors which were common to all or most of Scotland's coal-fired saltworks. Yet there were also some important differences between them. Changes in working methods too occurred over the long period with which this book is concerned, although rarely uniformly. Uneven development is perhaps one of the most striking of the technical features of Scottish saltmaking. In order to account for this a number of inter-related factors have to be considered, including proprietorial inertia and the degree to which individual saltmasters were concerned with efficiency; market pressures and opportunities; the force of local custom, and beliefs about best working practices and the variability and suitability of raw materials.

As was seen at the beginning of this chapter, the salt making process began with the collection and storage of a sufficient quantity of sea water. Filling the 'bucket pots' was a relatively simple matter, with water 'coming up with the flood tide through channels or watercourses'. Traditionally, each salt pan had its own bucket pot, from which water was transferred to the pans by hand in buckets attached to long wands or poles. This however was a slow and time-consuming process which was evidently being improved upon from the mid-seventeenth century, possibly in response to the still buoyant demand for salt exports (see Chapter Two). Thus in 1658, during the early stages of construction of Methil saltworks, it was considered best to 'make pumps to put the water...in the pans and not buckets as at Wemyss'.[65] Soon afterwards the hand bucket system at Wemyss must have been abandoned as, on his visit there in 1720, Henry Kalmeter noted that it was a 'windmill that draws up the water',[66] a method which was to be fairly widely used, particularly at saltworks elsewhere in Fife, in the eighteenth and early nineteenth centuries.[67] Horse-powered pumps were also used, as at Sir John Callendar's three-pan works in Prestonpans.[68].

Greater efficiency was also achieved at some of the bigger works by replacing individual bucket pots with a larger central reservoir which served a number of pans. Apart from allowing the better settlement of the 'heterogeneous ingredients' in sea water which so offended some contemporaries, larger tanks also went some way towards eliminating the irritation and expense involved in the upkeep and repair of several smaller bucket pots, and ensured water supplies for less-well situated pans. Thus the Wemyss windmill drew water into a tank which served eight pans; in 1772 a single tank was cut into rocks on the foreshore at St Monance, from whence water was drawn by a windmill into nine pans, in an arrangement which one observer described as the 'neatest and best contrived...upon the coast'.[69] From around the same time the 9th Earl of Dundonald's five pans at Culross were served by a massive reservoir, the ten and a half foot thick, twelve feet high walls of which enclosed a collecting area measuring 75 by 60 yards.[70]

At other works however hand pumping was carried on. This was understandable at small single pan operations such as that at Turnberry in Ayrshire,[71] but less so at works like Grange near Bo'ness where the water for the thirteen pans long continued to be pumped up 'by several men or women, often with young people helping them' until steam power was introduced, probably sometime in the early 1800s.[72] It is possible, but by no means certain, that the continuation of this practice, unknown on the Fife side of the Forth in the eighteenth century, reflected the greater tendency for saltmaking on the south bank of the river to depend more heavily on the family unit. What is clear however is that most works which survived the repeal of the protective salt duties between 1823 and 1825, made use of steam pumps and larger reservoirs, as at St Davids where sometime prior to 1836 a steam pump had been erected to fill a reservoir which could hold a week's supply of water for the pans.[73] As will be seen in Chapter Two, while most Scottish saltworks disappeared in the decades immediately following 1825, there remained sufficient demand for the product of those works which remained in business by refining rock salt shipped north from Liverpool, to justify the greater capital outlays on such equipment. At the same time, their survival in the more competitive market circumstances of the nineteenth century was to a great extent dependent upon continuous low cost production, which steam technology, in part, made possible.

Panhouse technique varied too. Some works were more efficient than others while there were also considerable differences in the quality of the salt. As has been pointed out, for most of the period the typical Scottish pan measured 18 by 9 feet. There were some exceptions however. Generally speaking, the greater the dimensions of the pan, the better was the quality of the salt. Larger grains were produced by slower evaporation. Concern with quality however, was rarely high on the established saltmasters' agenda. Indeed as will be seen in Chapter Four they made strenuous attempts to resist the introduction of improved methods. After the 1660s, when they had first obtained an element of protection from imported salt, the generality of Scottish saltmasters showed even less interest in the quality of their product. Few if any contracts with salters after this decade contained clauses by which Wemyss salters were bound, to deliver 'clean white salt', dried according to specific instructions.[74] Thereafter and until 1823 criticisms of the Scottish product became increasingly loud, as domestic consumers were unable to buy anything but salt which was often dirty and had a 'nauseous' flavour. As it tended to liquify, merchants buying in bulk usually preferred to be supplied with good 'old' salt, which had been allowed to drain for three months after its removal from the pans, although others were criticised for simply being concerned that they got 'good weight'.[75]

In the late seventeenth century however, and through the eighteenth, mercantilist philosophy, scientific interest and a yearning from some businessmen (who also had close links with the creative world of Scottish science during the Enlightenment) to break out of the market to which they were restricted by the generally poor quality of Scottish made salt, combined to inspire both the construction of new and bigger pans, and experiments designed to make better salt. By relieving the Scottish fishing industry of its dependence upon imported Bay salt, the establishment of a successful means of making fishery salt in Scotland would have been a major contribution to the Scottish economy.[76] While eminent scientists of the age, such as Professors William Cullen and Joseph Black, experimented in the field,[77] arguably the more practical schemes were those devised separately by Dr John Roebuck, better known for his involvement with vitriol and Carron ironworks, Archibald Cochrane, 9th Earl of Dundonald, and Dr Francis Swediaur, physician and entrepreneur. All three failed.

In the mid-1760s, 'the then great demand for salt' was sufficient to encourage Roebuck to build what was Scotland's biggest ever saltpan, the dimensions of which were 55 by 32 feet. Employing ten or more salters rather than the usual two, the Corbyhall pan (at Bo'ness) had been shut down by Roebuck's creditors by the end of 1771, partly owing to the technical problems which had been encountered in heating such an expanse of iron evenly.[78] So proud was Dundonald of his improved method of making salt, that in 1786 he invited Joseph Black and Adam Smith to come and inspect his 'lately erected' refining apparatus.[79] In spite of having built five pans of the larger Tyneside dimensions though, Dundonald's hopes were dashed by his inability to import rock salt freely from England. Like many others before him, Dundonald discovered just how weighty was the political influence of the Scottish marine salt interest at Westminster. Francis Swediaur's scheme of 1784 was the least developed of the three, and seems to have come to an early end as the bucket pot he constructed at Port Seton was evidently unable to contain the volume of water his pans required.[80] Paradoxically, Swediaur may have been better prepared, having lived in Holland and studied Dutch salt production methods, which he had hoped to introduce to Scotland.

To these grandiose attempts to depart from standard Scottish saltmaking practice should be added that of another prospective saltmaster with scientific connections, Arthur Martin of Kincaple, near St Andrews. In May 1758 he was reported to be building beside his existing pan an 'uncommon structure', of wood, in order to make an improved quality of salt. Two years later the works were in considerable difficulties, largely owing to the problem of getting regular shipments of coal brought to St Andrews and thence by horse to Kincaple. In 1766 his pans were finally abandoned.[81]

Several factors accounted for the variable quality of common salt. At Leven for example it was reckoned that the salt was 'not so salty by about one third' as at neighbouring Methil, owing to the pans' proximity to the river Leven. The water (at the pans) 'was not so briney' as it was at neighbouring Methil.[82] Wisely, when building his new set of pans at Methil in 1658, the 2nd Earl of Wemyss had decided to site them west of the harbour 'furthest from the water of Leven'.[83] Apart from quality, the costs of production were also raised at saltworks which were sited close to rivers. With no nearby flow of fresh water of any consequence, at Saltcoats the ratio of sea water to salt was 25:1. The norm on the Forth was 28:1, but at some works the ratio could be 32:1, which made it necessary to refill the pans with water five rather than four times, and in consequence to burn more coal.[84] Even so, salt from west coast pans was considered inferior to that from the Forth. It was said by one observer to have had a 'greyish unsightly appearance' compared to the lighter, larger grained salt made at east coast pans.[85]

THE SALTER'S CRAFT

While it is difficult to say why this was the case, one factor which is likely to have played a part was the degree of skill of the salters. Judged by some historians to have been a brutish, unskilled lumpen proletariat,[86] in fact they shared with coal miners of the period certain positive attributes which were not easily taught.[87] Like colliers they required strength and stamina, to keep the pans going and draw off the salt. In addition though, the job involved certain 'knacks' which could only be acquired through lengthy familiarity with the salt making process. Pans were easily burnt, while inadequate control of the heat made for poor wet salt if evaporation was allowed to take place too quickly. Thus in an account of the former salters at Grange, H M Cadell recognised that there was 'an art here which outsiders could not readily acquire'.[88]

This was certainly a view which would have been shared by one of the proprietors of the struggling Maryburgh saltworks at Prestwick, who was urged by the works' manager to ensure that he had good salters, as the job required 'great care & attention, as well as skill'; to bring the pan 'to Salt' was 'a nice operation'.[89] With its long and extensive salt making traditions panmasters on the Forth are likely to have been relatively well-endowed with able salters, especially during the eighteenth century (see Chapter Six). In Ayrshire, saltworks were thinly scattered, while the industry was extremely small prior to the major developments in coal and salt which were undertaken by Sir Robert Cunninghame of Auchenharvie in the 1670s.[90] As a result the acquisition and retention of salters was a recurrent problem, and 'East coast men' were much sought after. During the construction and early operation of the saltpan at Cock, on the island of Arran for instance, in 1710 and 1711, the salters had to be brought over from Bo'ness, at some cost, as they could not be recruited locally.[91]

Yet it would be a mistake to draw too firm an east-west division. Unsatisfactory salters were also to be found at pans in Fife and the Lothians. In the autumn of 1730 for example, an unsuccessful attempt was made to entice the master salter John Hound from the Earl of Wemyss' pans to Wemyss of Bogie's works, 'to supply Mark Muir's place who had run in arrears and was unskillfull'.[92] At Grange in 1779 George Robertson

4 Stirring the brine, Prestonpans saltworks, 1958. Little had changed, either in terms of the construction of the saltwork or the technique used, since the seventeenth and eighteenth centuries. During the long evaporation period (24 hours or more) the salter and his assistant drew off impurities which rose to the surface and tried to ensure that the brine evaporated evenly around the pan. The boxes in the left foreground are 'drabs' into which the salt was put after having been drawn from the pan. Photograph by S. F. Sanderson, *School of Scottish Studies, University of Edinburgh.*

was replaced by James Stanners, who was evidently incapable of making the weekly quota of four draughts of salt.[93] As will be shown in Chapter Five however, it was not only the extent of their skill and physical condition which determined the salters' contribution to the fortunes of their respective works, but also their attitudes to work and authority, both as individuals and collectively.

Although most salt pans in Scotland were of a similar size, firing arrangements could vary quite substantially. The 'sole' pan method was that most commonly used until the later eighteenth century. It has been well described by Kalmeter, after his visit to the pans at Wemyss in 1720:

> The wall in which the pan is set has an opening underneath which is just as wide as the pan, exactly like the arch of a bridge. Here they throw in the...small coals, and heat the pan with their flames.[94]

How far the pan was from the floor below, and hence the fire, seems to have been heavily dependent upon local tradition. At Cockenzie in the 1720s for instance there was five feet between the pan base and the floor. Making by this means, 'such a quantitie of Salt as...no pan north side of the water ever makes', it was judged far superior to the practice adopted in Fife, where only two feet of space separated the floor from the pans, which could not be heated evenly, and were often badly scorched.[95] The most efficient sort of pan however was the 'brander'.

Rather than being simply thrown onto the floor, coals were placed on a piece of perforated iron which hung beneath the pan.[96] Although 'brander' pans had been known in Scotland at least as early as 1720, when they were observed at Leven, they were not widely used until much later in the century. Their early introduction at Leven may have been due to the influence of Daniel Peck, who leased the saltworks there in 1719. An English merchant adventurer from Cheshire with numerous business interests, Peck had in 1706 patented an improved means of making and refining white salt.[97] The lag in the adoption of brander pans may possibly be explained by the relatively slack demand which existed for Scottish salt for most of the 1700s, which made unnecessary the greater financial outlay required to construct pans which could increase salt output by at least 25 per cent. Even though they became more common during the last years of the eighteenth century, they were not universally adopted. At Dysart, for example, in 1817, after several years of intense demand for salt, only one of the six pans was of the brander type.[98]

PRESSURES FOR CHANGE

What did inspire the attention of most saltmasters from the later eighteenth century however was the need to reduce operating costs. While fuel efficiency had always been considered, it was only after the 1760s that this became a more pressing matter. Beforehand, demand for 'small' coal had generally been weak, and accordingly fuel costs had been relatively low. As industrial and agricultural demand for small coal grew however, its cost

5 Salters rake, 1958. This long-handled implement (seen in use in plate 4) was used to stir the brine and and draw salt to the pan sides prior to its removal. Photograph by S. F. Sanderson, *School of Scottish Studies, University of Edinburgh.*

rose,[99] and, along with the more competitive circumstances of the salt trade generally, proprietors were forced to rationalise production methods and alter long-established ways of doing things. 'Steadiness and economy', the partners of Grange saltworks agreed in September 1792, would henceforward guide their operations. There and at other works, contracts with salters and other workers were redrawn, while more efficient working methods were sought. At Pinkie the outcome appears to have been a substantial reduction in the coal-salt ratio, to an annual average of 3.17:1 between 1784 and 1797.[100] From the early nineteenth century, efforts were made to utilise the waste products from salt manufacture, notably the formerly discarded 'bittern' which was used at Saltcoats to make Epsom or Glauber salts.[101] Elsewhere schemes were proposed which would, if successful, have halved the time it took to make a full pan of salt and tripled output with constant costs.[102]

Environmentalist pressures too played their part in effecting technical and operational change. Hugh Currie, for instance, who in 1814 leased some ground from the Eglinton estate in the planned and partly built new spa and port town of Ardrossan, was bound to construct the panhouse furnaces and vents 'in such a way and manner as to condense their own smoke...so it may not be a nuisance to the public'.[103] Neither Currie however, who signed a 36-year lease, nor any other Scottish saltmaster, had access to technology or cost-efficient working methods effective enough to counter the power of free competition in Scotland with Liverpool salt, with which they were faced, with generally devastating results, after 1825.

2

OUTPUT, SALES AND MARKETS

c. 1570–1707

By the second half of the sixteenth century salt manufacturing was an established if small Scottish industry, having been carried on at several locations around Scotland's coastline for a considerable time. By the beginning of the 1570s substantial numbers of pans had been erected, with the heaviest concentrations of these being found in the vicinty of Prestonpans and Musselburgh, where there were 31 pans, followed by Kirkcaldy with 28, Dysart with 16, while last but not the least significant was the group of 11 shared between Culross and Fordell.[1] Although little is known prior to this period either about the industry's fortunes or about the quantity of salt made in Scotland, it does appear that a modest export trade from the river Forth was conducted from at least the 1460s, with Dysart and Prestonpans for example playing a prominent role from the 1470s.[2] Shipments had gone to Norway and England, although by the 1560s almost eighty per cent of Scotland's exported salt was going to the Baltic.[3] Exceptionally low saline levels in the Baltic Sea, which precluded the successful manufacture of sea salt, made that region critically dependent upon external suppliers.[4] It is clear too that some Scottish saltwork proprietors and their factors were alert to the market opportunities which were provided by interruptions in the European salt trade and short-term relative price changes. Such an opportunity, to achieve sales in Danzig and perhaps further afield, was evidently seized between 1474 and 1490 when the combination of a shortage of Bay salt and high prices created the necessary conditions for what may have been the first of two 'mini' export booms for the Scots which occurred prior to the marked and sustained upturn of the 1570s.[5] Between the late 1470s and 1486 the quantity of salt shipped from Dysart doubled.[6]

By this time the European market environment had undergone a long period of change which was highly favourable to the comparatively high cost salt makers of Scotland and Tyneside. Demand was rising as, in general, the sixteenth century was an era of demographic growth in western Europe. Supply-side conditions however were far from secure. From the fourteenth century the salt makers of the Low Countries, with their highly esteemed product, had been experiencing growing difficulties as supplies of their unique raw material, 'zelle' or salt-impregnated peat, diminished.[7] From the early fifteenth century another formerly important and high-quality European supplier, the brine salt industry of Luneburg, also began to decline.[8] In addition the scale of England's coastal salt industry in Lincolnshire was reduced, partly because entrepreneurial interest shifted towards the more lucrative trade in woollen cloth, partly as wage costs rose.[9] Fortuitously, although also contributing to the demise of the north European salt industries, ample supplies of cheaper and qualitatively sufficient Bay salt could be obtained from the salines on the Atlantic coasts of western France and Iberia.

From the late fourteenth century onwards the seaborne trade in Bay or 'great' salt grew, with vast quantities being shipped northwards to Scandinavia, the Baltic, the Low Countries and the British Isles.[10] This trade however was highly vulnerable. Political disturbances within the salt producing regions or the outbreak of hostilities involving the nations supplying or shipping Bay salt could and did lead to periods of acute salt shortage.

Religious and political disturbances in France in the mid-sixteenth century, which interrupted salt production and caused prices to rise, evidently revealed in Scotland as well as England the dangers of depending too heavily upon a single and distant source of salt.[11] Inflated prices drained scarce currency abroad while a shortfall of imported salt obviously had deleterious consequences for the preservation and export of fish. Salt shortages and high prices produced considerable hardship for domestic consumers in Scotland while failure by the authorities to ensure adequate supplies at reasonable prices engendered popular hostility and gave to their opponents a legitimate and politically potent cause of complaint.[12] For the crown then there was much to be gained by the expansion of the native salt industry, especially if sufficient improvement in the quality of the Scottish product could be achieved to enable it to serve as an adequate substitute for the preferred Bay varieties. Thus in the early 1560s, north as well as south of the border, attempts were made to persuade foreigners to introduce and establish their superior salt making methods.[13] None succeeded. Angelo Manelio for instance, the Italian salt maker encouraged by Mary, Queen of Scots, was reported to be in disgrace by 1565, two years after his arrival.[14] A subsequent scheme for making the prized 'great' salt, which involved Anthony Hickman, a Londoner, and Francisco Berti, a Florentine, surfaces only briefly amongst surviving historical records.[15]

It was not official encouragement which accounted for the pace and pattern of the salt industry's development in the late sixteenth century. This was determined by private capitalist enterprise from the landed and mercantile sectors in response to exceptionally promising market opportunities overseas. The principal carriers of Bay salt were the Dutch, whose salt-laden vessels supplied not only the needs of their great North Sea herring fishery but also, after the Bay salt in its crude state had been refined, the substantial Baltic market.[16] In 1566 however the so-called Dutch Revolt began, a colonial conflict between Spain and the Low Countries which was to involve lengthy periods of warfare between the two nations. For the Spanish an obvious and potentially devastating weapon to use against a state so dependent upon maritime trade as Holland was to cripple the latter's seaborne trade with Iberia, a substantial part of which comprised salt.[17] It was then the existence of Dutch-Spanish conflict and the waging by Spain of intermittent economic warfare throughout the period of a rebellion which was to last until 1648, which provided the high-cost, low-quality salt makers of Scotland with unprecedented opportunities to expand their sales overseas, not only in the Low Countries but also to temporarily salt-starved customers of the Dutch. The general shortfall in salt supplies which Spanish actions created caused English consumers too to search for Bay salt substitutes, which inspired a substantial increase in the size of the sea salt industry on Tyneside.[18]

Although demand-side circumstances generally were entering a favourable phase, they were by no means static. Openings for additional sales often appeared suddenly and could disappear almost as rapidly as Spanish embargoes on Dutch trade were imposed, in 1572, 1586, 1595, 1599, 1601 and 1603 for example, and lifted.[19] It is important not to exaggerate the extent to which shipments of salt into the Baltic were disturbed. Thus while there were violent short-term fluctuations in the quantities of Portugese salt shipped through the Danish Sound, in the century or thereabouts following 1562 annual average shipments never fell below 22,000 lasts and never rose above 33,000.[20] Scottish saltmasters and merchants then had to be ready to strike fast and hard when bigger than usual gaps in markets abroad appeared.

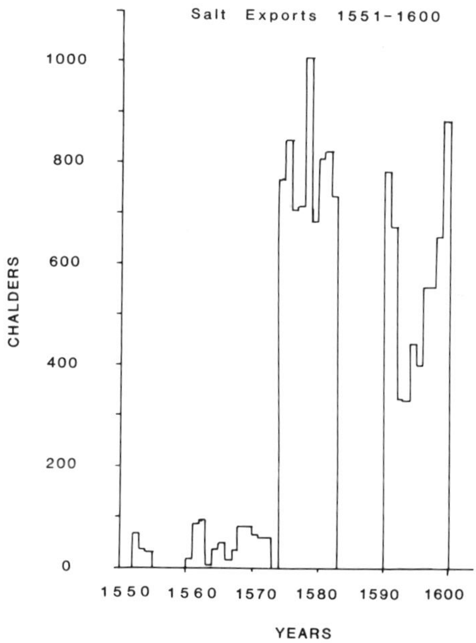

Figure 1 Scottish salt exports, 1551–1600.
Source: adapted from I. Guy, 'The Scottish Export Trade, 1460–1599', unpublished M. Phil. thesis, University of St. Andrews, 1982.

The immediate response to the first Spanish embargo in 1572, when contacts between North Netherlands merchants and Spain were rapidly ended,[21] and which resulted in a quadrupling of the price of Bay salt,[22] was to utilise existing salt making capacity to the full and divert as much salt as possible from the Scottish market. From under 100 chalders per annum at the turn of the 1570s, as Figure 1 shows, exports of Scottish salt rose steeply to over 800 chalders in 1574 and remained around and occasionally above this level until the early 1580s.[23] While there is some uncertainty as to the exact distribution of this eight-fold increase in exports, it is clear that the principal markets were the Baltic, the Low Countries and England. Previously the main export market however, the Baltic's relative importance was substantially diminished, and during the peak years of the boom, 1574–78, its share of Scottish exports fell to well under 20 per cent even though the quantity of Scottish salt going there was three times the level it had been during the previous decade.[24] Most seems to have been shipped to Holland, although salt also formed a substantial portion of Scottish trade to England, with shipments going to English east coast ports such as Hull, Ipswich and London in the 1580s and 1590s.[25] A further indication of Scottish success in this area comes from the complaints of the salt makers of Lincolnshire about competition from the 'good chepe' salt of Scotland.[26] Although as will be seen later, England was probably only briefly as important a market for Scottish salt as some historians have suggested, links established this early with the south-eastern English fishing trade may account in part for the longevity of this outlet for what was generally only a small proportion of the output of some of Scotland's saltworks. The trade however was to continue in evidence well into the eighteenth century.

Impressive as the rise in Scottish salt exports was, the Scots' contribution to the European salt trade as a whole was negligible. Although available figures are approximate and precise comparisons should therefore be treated with caution it is instructive to note that during the peak years of the 1570s the annual average quantity of Scottish salt shipped into the Baltic amounted to less than one per cent of the lowest annual average figure for Portugese salt alone shipped there.[27] In stark contrast the impact of the Scottish salt masters frenzied overseas adventure was within Scotland itself profound, and indeed had consequences of which there was still a faint echo to be heard more than two centuries later, in 1794.[28] By the spring of 1573, Parliament, under the Regent the Earl of Morton, who was concerned to limit the export of several basic commodities,[29] was evidently becoming alarmed about the high prices of small salt. A three year ban on exports was imposed, although salt could still be shipped to Norway, a concession designed to ensure the continuation of timber imports.[30] The ban was not total though and John Blaw and Alexander Ysett, for themselves as well as on behalf of the others in Culross and Fordel, were amongst the first of many saltpan proprietors who undertook not to export salt, 'unto sic tyme the subjectis be sufficientlie furneist'.[31] They agreed too that salt should be sold from the pans at a fixed price, 12 shillings Scots for the larger Culross boll and 8 shillings Scots for the more common measure. The potential advantage to the exchequer of permitting salt to be exported was not overlooked and for each chalder of salt which was shipped overseas, six ounces of silver, nine for the Culross equivalent, was to be paid to the bullion-hungry authorities.[32] Many such undertakings, which in spite of intense efforts by the authorities, were not given by every saltmaster, were evidently largely ignored however, with exports being maintained at record levels through the mid-1570s until 1582.[33] Complaints about a 'derth of small salt within this realme' continued to be heard as the moral element of what remained of medieval economics was cast aside in the cause of 'privat lucre'.[34]

Official concern about salt supplies appears to have abated but not disappeared from 1575 and this may indicate some amelioration of earlier difficulties. Two factors may account for this. First, and perhaps the least convincing, was the Privy Council's decision, in December 1574, to play a direct role in the distribution of domestic salt supplies.[35] Precisely how, how well and for how long this scheme operated is unclear. Secondly, and of far greater consequence, was the fairly considerable expansion of the industry's productive capacity which the boom conditions had encouraged. By the end of 1574 it had been recognised that Scotland's 'pan maisteris' had 'biggit mony ma salt pannis and set mony ma servandis to werk' than ever before. Undoubtedly the most striking advance was made in and around Culross under the direction of Sir George Bruce who from 1572 and prior to 1625 marshalled a veritable empire of forty-four pans.[36] Previously Culross had boasted only seven saltpans.[37] This development seems to have considerably eased the pressure of demand for salt in Scotland although there continued to be short periods of hardship when not only war but also a spell of bad weather could cut off supplies from Iberia or Brittany.

State interest in the salt industry itself, however, was maintained with the revenue-enhancing possibilities of a successful export trade tending to replace earlier concern about domestic supplies. From 1584 export duty of 3*s*. 4*d*. Scots per chalder was charged, to be used solely for purchasing ordnance for the crown as well as to provide a repair fund for 'ruynit' royal 'castellis and houssis' which 'aucht yeirlie to be reparit, mendit and augmentit'[38]; John Chisholme, comptroller of ordnance, managed the salt duties for this purpose until the end of the century.[39] Schemes to improve the quality of Scottish salt also continued to receive encouragement. In 1587 Dame Margaret Balfour in Pittenweem was granted a seven year right to make 'salt upon salt' on the grounds that many sorts of fish could not be salted with 'the small salt qlh is accustomit to be maid or hes bene maid befoir within this realme'. 'Gryt salt' from 'hispanie' and Brittany was expensive and subject to 'perrellous viages' and other difficulties.[40] A year later Eustachius Roghe or Roche, a Flemish 'medicinar' who from 1583 had been granted extensive mining rights in Scotland, extended his interests into salt, of which he proposed to make vast quantities well and cheaply, thereby increasing state revenues by 100,000 merks.[41]

While no direct evidence survives which would make it possible to evaluate the success of these schemes, that other attempts to raise the

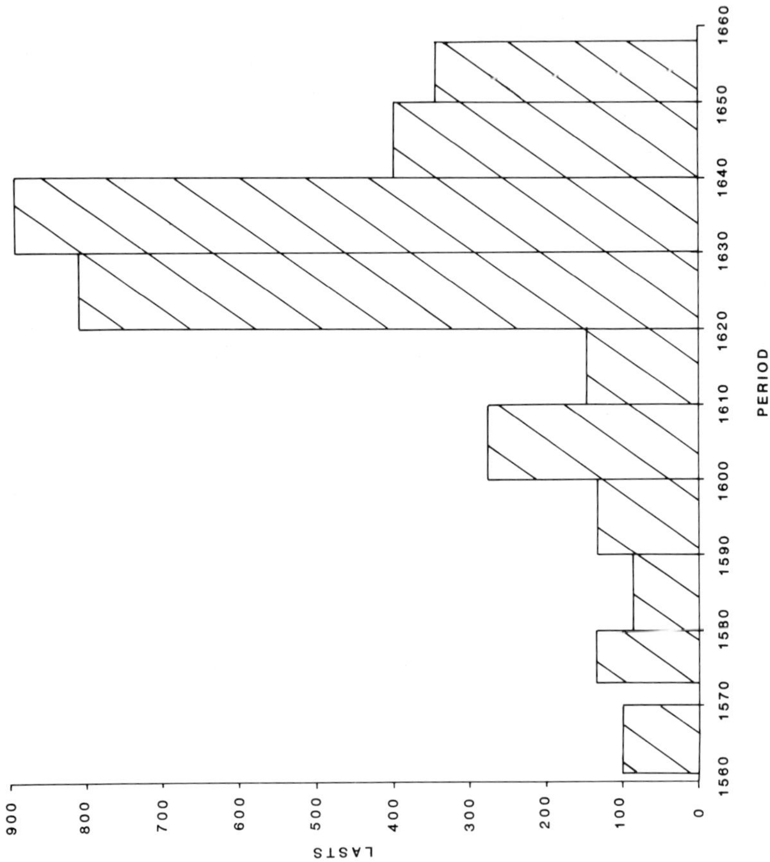

Figure 2 Scottish salt entering the Baltic, 1560–1660.
Source: based on data from S. G. E. Lythe, 'Scottish Trade with the Baltic, 1550–1660', in J. K. Eastham (ed), *Economic Essays* (Dundee 1955), 74.

standard of the Scottish product followed, as in the cases of Robert Monteith in 1621[42] and Nathanial Udward in 1630,[43] suggests that they had little lasting impact.

Although there were marked fluctuations, conditions for sales abroad generally remained favourable in the later years of the sixteenth century and early decades of the seventeenth century. New saltpan construction, at Inverkeithing, at Brora in Sutherland in 1598 and 1614, and at Limekilns around the same date, is indicative of optimism from within the industry about the future.[44] What statistical material there is available is used graphically in Figure 2 and shows that shipments of salt from Scotland to the Baltic were higher on average in the first two decades of the seventeenth century than they had been in the 1570s. Between 1600 and 1609 the annual average figure of some 259 lasts was more than double the previous best.[45] The relative scarcity of data and known interruptions in the trade however make it unwise to talk in terms of long term trends. In 1599–1600 London received 504 weys of Scottish salt; in 1620–21, 1,336 and five years later only 575 weys.[46] Other English ports also imported salt from Scotland. Notable amongst these were places such as King's Lynn and Boston,[47] of which Sir George Bruce's pans were major suppliers, a position which was undoubtedly assisted by the monopoly of this trade he was apparently granted by King James VI in 1611.[48] It was northern Europe and Holland in particular though which were the main customers for Scottish salt in this period.[49] What is equally clear is that marine salt was a leading Scottish export, with the 2,000 chalders which were shipped abroad in 1614 then exceeding the value of coal exports.[50]

The peak period for the export of Scottish salt in the seventeenth century—indeed for all time—was the 1620s and 1630s. The end of the Twelve Years Truce between Spain and the Netherlands in 1621 began a period of even more determined economic warfare by the Spaniards, reinforced by naval activity designed to disrupt Dutch merchant shipping and fishing fleets.[51] In consequence the Dutch-Iberian trade was reduced to a mere trickle, with little Spanish or Portugese salt arriving in Holland between 1621 and 1641.[52] The potential of this situation for the Scots increased as the price of salt moved sharply upwards during the 1620s. In Scotland then, as on Tyneside, output was increased and a wave of new saltpan building activity was inaugurated. One of the first to respond was David, 2nd Earl of Wemyss, who in 1628, added the first of at least two pans to his existing saltwork at Wemyss.[53] A further spur was added in 1630 when exports of French salt to Scotland were checked[54] and in the course of the next three years, which overlapped with the period in the Spanish-Dutch conflict when salt prices in Holland were almost at their highest,[55] at least three new Scottish saltworks were established. Of greatest long-term significance were the works at Cockenzie, under the direction of George, 3rd Earl of Winton, and Joppa in Duddingston parish, which belonged to David Preston of Whitehill and Craigmillar.[56] Underlining the extent to which prices had risen and opportunities for sales overseas had opened up was the construction in 1633 of twelve pans, 'which for some time were all at Work' by John Stewart, Earl of Carrick, on the Orkney island of Eday, where as was seen in the previous chapter, the absence of coal necessitated the use of peat for firing purposes.[57]

The principal markets, as has been noted already, were Holland, to which Sir William Brereton reported in 1636, 'the greatest part' of Scottish salt exports went, and the Baltic, although the English outlet too was maintained. On average almost six times as much salt was sent to the Baltic in the 1620s and 1630s than in the previous decade.[58] The boom however ended abruptly. Changing political circumstances in mainland Europe, notably the Portugese Revolt against the Spaniards in 1640, burst the Baltic and Low Countries 'bubbles' for the Scots and by the spring of 1641 the Dutch had managed to regain control of the international salt carrying trade.[59] So dependent had the Scots apparently become on the Dutch and Baltic markets however, that the cessation of hostilities between Holland and Spain, which were ended with the Peace of Munster in 1648, might reasonably have been expected to have caused the elimination of Scotland's salt from the international marketplace and as a result set the industry on a course of inexorable decline.

Yet this did not occur. Most of the evidence from within the industry from around 1640 until the early eighteenth century points not to a lengthy process of decay but instead to survival and at times growth and modest prosperity. Some works were abandoned, but as in the case of all but two of the Eday panhouses in Orkney, often stood as crumbling monuments to their promoters' over-optimism and do not appear to have reflected the condition of the industry in general. As was pointed out in the previous chapter, the pans on the coast at Innerwick near the East Lothian village of Dirleton, for example, were deserted in the 1680s for no other reason than that the colliery which had supplied them with fuel had been worked out and was flooded.[60] Yet in Fife there were no obvious signs of decay. In the 1640s the presbyteries of Kirkcaldy and Tulliallan made frequent attempts to stop Sunday working at the pans, an opportunity for rest which was far from graciously accepted either by their owners or the salters. It was apparently only with the greatest reluctance that the pans at Wemyss ceased production on Sundays, with the Earl of Wemyss and two neighbouring panmasters, Sir George Hamilton and Sir John Wemyss of Bogie, submitting more than one paper to Kirkcaldy presbytery, 'importing the necessities of the goeing of Saltpans upon the Sabbath'.[61] As late as 1650 John Archibald, an elder of Tulliallan presbytery, was 'exhorted in the name of Chryst to mak more conscience of visiting the saltpans within his bounds', and to deal with those 'who do not put furth their fyre on Saturdayes night'.[62]

While caution should be exercised about generalising from the reasonably well documented activity of the Wemyss family, that the 2nd Earl felt confident enough about the future in 1659 to enter into a five year contract with Robert Hunter, a smith, to maintain the twelve pans at West Wemyss, makes less convincing the complaints of some Scottish saltmasters in the early 1660s about their trading difficulties.[63] Such pleas, as will be seen in

Chapter Four, were used rather liberally to justify their claim to a monopoly of part of the Scottish market. Even more striking was Wemyss' decision to extend his saltmaking capacity by erecting a range of pans near his new coal port of Methil. By the 15th of July 1665 smith work on the first pan was completed.[64] Expansion at Methil was followed elsewhere in Fife. In 1693 for instance the burgh authorities of Inverkeithing let a piece of ground to David Maither, a smith, 'for his building saltpans and buckit potts', the number of which was increased by Colonel John Forbes of Pittencreiff around 1705.[65] Important developments took place elsewhere in Scotland too. In the 1670s for example, salt began to be made near Campbeltown on the Kintyre lands belonging to the dukes of Argyll, while shortly afterwards a marked expansion of saltmaking capacity took place on the Ayrshire coast at Saltcoats, under the direction of Sir Robert Cunninghame of Auchenharvie. Further west at least two new saltworks were opened on the island of Islay between 1692 and 1696 while further Hebridean expansion followed with the construction of a pan on Skye by Magnus Prince in 1703.[66] Salt was first made on the Hamilton family-owned island of Arran in 1710.[67]

How is this apparent contradiction between collapsing overseas markets and continued investment, often in parts of Scotland which were far removed from the established trading links and rich coal fuel reserves of the Forth, to be accounted for? Sales to Holland did plummet and as in the formerly useful Norwegian market, were of little consequence after 1660.[68] Nevertheless the Forth saltworks still managed to obtain sales in the Baltic and while annual average shipments of some 400 lasts in the 1640s, and a little less in the 1650s, were less than half the level of the spectacular 1630s, this was still twice as much as was sent there, on average, in the first two decades of the century.[69] Baltic sales held up at these lower levels (but which on occasion could reach 85 per cent of the 1630s peaks)[70] until around 1700.

The potentially devastating blow of the loss of the Dutch market however was softened principally by an expansion of shipments to England, which until the end of the 1660s may have been the industry's single largest market outside Scotland. Around 1670 one contemporary estimated that 28,000 large bolls of Scottish salt were going to England, compared to a total of 22,000 sent to all other markets outside Scotland.[71] In the year ending September 1662 English ports received a total of 2,828 weys of salt from Scotland,[72] which represented what was almost a five-fold increase over the level of the mid-1620s and was a considerably higher annual amount than ever went to the Baltic, even in the peak years. During the period of the Cromwellian union and for over a decade thereafter the Scots were able to compete successfully with the Tyneside saltmakers, apparently mainly because of lower production and freight costs.[73] While the Restoration saw the imposition of higher import charges for the Scottish product, these were not sufficient to cancel out the Scottish producers' advantage over their Tyneside rivals and as late as 1668–69 1,715 weys, or 61 per cent of the 1662 figure, went south.[74] Further pressure on the English Parliament on behalf of the Tyneside salt interest drove the tariff on Scottish salt up to the prohibitive level of 18*s*. 8*d*. per wey and by 1680 shipments had all but disappeared.[75] On top of this catalogue of disappearing overseas outlets for Scottish salt was the discovery, in 1670, of rock salt in Cheshire, while from the early seventeenth century the Northwich salt boilers had been improving their production methods and raising output. Ominously for both the saltmasters of north-east England and the Forth, in 1706 several Danish vessels arrived at Liverpool in search of cargoes of salt.[76]

Yet the fortunes of the salt industry cannot be explained solely by reference to its performance in markets abroad, although this is what most historians have done in the past. As has been pointed out earlier, Scottish salt users had a preference for French, Portugese and Spanish Bay salt and throughout the sixteenth and seventeenth centuries this was imported to Scotland in vast quantities. Nevertheless, although the evidence is scanty and altogether unquantifiable, it does appear that there had long been a fairly significant sale of Scottish salt in the home market, notably to the inhabitants of those areas within the vicinity of the saltworks of Fife and the Lothians, but also further afield. It was presumably sales within Scotland which had sustained the many pans belonging to the monarch and the church in medieval times. Home sales too would account for the survival of the industry through the fifteenth and sixteenth centuries prior to the great expansion of exports in the 1570s. Even after the export sector had become firmly established home sales continued to consume part of the industry's output. Thus in the early seventeenth century one visitor to Fife noted that there was made 'an immense quantity of the whitest salt...which, in addition to the local consumption is sent away for sale to outsiders'.[77] In 1618 at least part of the 90–100 tons per week which it was claimed were made by Sir George Bruce at Culross was disposed of within Scotland[78] while Thomas Tucker's 'Report' of 1656 points to the existence of at least a modest inland trade, with salt being sold from the pans 'to cageors and other poor people, who carry the same about in creiles on horseback, or otherwise, up and downe the countrey, for the expence and consumption thereof'.[79] Such traffic was likely to have increased as the Biscay salt makers ran into difficulties around 1600.[80]

With the strategic importance of the home market growing, the saltmasters, forced out of formerly lucrative markets and facing from the late 1650s onwards the threat of being shut out of England, made determined and ultimately successful attempts to dispose of a much larger proportion of their salt in Scotland. Large tracts of the country, especially in the west and north-east, still depended almost exclusively on imported salt. While much of this was necessarily used for fish and meat preservation, it was the large quantities imported for domestic use which the salt industry's leaders sought to substitute with their own inferior but adequate salt. Chapter Four will look at the concerted activities of the saltmasters in greater detail. It is sufficient here to establish that by the end of 1665 the Scottish legislature had become firmly committed to a policy of protection for the salt industry, with an additional £12 Scots being imposed on each 'watter boll' of foreign salt. Only salt which was to be used for preserving fish and meat for export could enter Scotland free of duty.

Thus the impact of the loss of the most important remaining market outside Scotland, England, was compensated for by the provision of the legal conditions for a monopoly of the home market. Consumer resistance and the inefficiency of customs officials of course slowed the rate at which substitution of Bay salt by the Scottish variety took place. Nevertheless during the closing decades of the seventeenth century Scottish salt does appear to have increased its market share. Around 1670 it was estimated that some 40 per cent of the industry's output was sold in the home market; by 1706 this may have risen to just below 70 per cent.[81] The trend which is revealed by these rather impressionistic sources is confirmed by the far more precise Salt Office data, which shows that in the period 1716–9 three-quarters of Scotland's salt was sold within the country of its manufacture.[82] On average, home sales accounted for around 67 per cent of total recorded sales of Scottish salt in the years between 1714 and 1720 for which sufficient data exists (see Table 2.1).

While this goes a long way towards accounting for the visible buoyancy of the industry in the two or three decades prior to the Union of 1707, by itself it is not sufficient. War continued to play an important part in determining the industry's fortunes. This was true for overseas sales, mainly in the Baltic by this time,[83] as well as in Scotland. Their dependence upon such extraneous factors was clearly recognised by at least some of the industry's leading entrepreneurs. Thus Daniel Hamilton, reflecting upon the 'want of saltpans' on the Kinneil estate of the dukes of Hamilton in 1696, was concerned about 'the uncertainty of the counteanouance of the warrs upon which the profit Depends for how soon we have peace with france the price of coalls and Salt will fall' and considered that further investment would be risky.[84] Indeed it was the Anglo-French war, waged almost continuously from 1689 until 1713 and which cut off normal supplies of French and Spanish salt, that provided the impetus for the additional saltpan construction which occurred in Scotland in the 1690s and during the early years of the eighteenth century. The Scottish industry was not the sole beneficiary; in Cheshire too output rose, while it was during this period that Newcastle's salt trade neared its peak.[85]

In national terms at least, the Scottish demographic disaster of the 1690s, which may have reduced the population by anything between 5 and 15 per cent or even more,[86] seems not to have had an adverse effect on the industry. This is probably because such was the shortfall of Bay salt that however reluctantly, fish curers and others were temporarily forced to utilise the native product. Thus by the late 1690s unprecedentedly large quantities of Scottish salt from Kirkcaldy, Methil, Wemyss and elsewhere were being shipped into Dundee.[87] That Alexander Williamson, a Kirkcaldy merchant, purchased over 596 pounds Scots worth of salt from Sir James Wemyss of Bogie in 1702 'for salting and curing of Herrings'[88] is further testimony to the shortage of foreign salt. It is this which explains why pans were being built on Hebridean islands such as Islay and Skye, which although lacking in coal, had the advantage of being situated close by the west coast fishing grounds.

There were additional temporary openings in traditional markets too. In March 1698 for example Sir John Lowther of Whitehaven, keeping a close watch on the Baltic market, observed that 'French salt is so dear that they bid them send Scotch salt; coals and grindstones are also certain'.[89] These decades also saw a further flurry of interest in improving Scottish (and English) salt manufacturing methods, although such schemes were generally opposed by the established saltmasters of the Forth who were content, indeed anxious, to maintain their share of the Scottish market by supplying inferior salt made by traditional methods. However the growing restrictions upon the use of unrefined Scottish salt for meat and fish for export, the continued shortfall of Bay salt and the imposition by Parliament in 1706 of a ban on the importation of salt from both France and England, as political temperatures rose, provided an environment conducive to those prepared to indulge in chemical experimentation.[90]

*c.*1707–1850

Current historical thinking about what happened to the Scottish salt industry after the Union of 1707 is far from uniform. Optimistic interpretations are in the minority[91] and the most commonly held view is that salt making in the eighteenth century was in decline, 'adversely hit by the union and the new fiscal system' which exposed it to competition from the cheaper and superior salt brought north from Liverpool.[92] Between 1780 and 1840 according to another historian, the industry was virtually wiped out.[93] There are two main explanations for this pessimism. First, too much emphasis has been placed on the industry's export performance, and, as has been the case for the seventeenth century, domestic sales have largely been overlooked. Secondly, judgements have been based almost entirely on impressionistic and non-quantitative evidence. Indeed it is the selective use of the latter which has been largely responsible for the variety of opinion on the subject.

Fortunately for the historian, but much to the disgust of most eighteenth century Scots, salt duty was imposed in Scotland in 1713 on terms settled in the eighth Article of the Treaty of Union. The vast bulk of the several thousand 'Salt Vouchers' upon which were recorded the quantities of salt sold and duty paid at each of the country's saltworks have survived for much of the period prior to 1823 when the salt taxes were repealed, thereby making it possible to calculate the total salt made in Scotland in most of the years when duty was payable. Further, owing to the differential rates of duty charged and a highly complex system of drawbacks and allowances, the officers responsible for collecting the duties were required to list the various categories of salt sales under separate headings, so allowing some disaggregation of the total figures. The most important sub-divisions were 'Retail', which was salt sold direct from the pans, 'Coastal', that is salt shipped from the pans to other parts of Scotland, and the self-explanatory heading 'Export'. Salt used for preserving certain types of fish for export was also recorded separately. Unfortunately it is only possible to disaggregate the data in this way until 1798 as thereafter the individual vouchers are not available. Nevertheless the series for total sales can be continued through to 1823.

Objections can be raised to the use of fiscal data to plot the course of the industry's progress over time. Above all there is the problem of tax evasion and consequent under-recording of real output. Such fears however can be largely assuaged in this case. Briefly, while there was a great deal of clandestine dealing in salt, by far the greatest part of this concerned the more popular imported varieties. Bay salt was being diverted from its legal (and duty-free) use in fish and meat preservation while increasingly from the mid-eighteenth century salt legally shipped from Liverpool to Ireland was being sent across the Irish Sea and illegally imported into south west Scotland.[94] It is probably true to say that illicit dealing in Scottish salt was endemic in the vicinity of the saltworks. Yet all of the available evidence points to an active market in extremely small individual quantities of salt, with by far the most common seizures of 'run' salt by customs officers being of one or two bushels, or what could easily be carried in a sack. The total quantities involved would not have made a significant difference to the recorded sales figures.[95] The incidence of large scale dealing at the pans was fairly slight, although there are occasional dramatic exceptions. Inevitably then the vouchers under-record the real quantities of salt which were made in Scotland.[96] Nevertheless it seems entirely reasonable, if due caution is exercised, to consider the trends which the series based on this data reveal as valid indicators of the industry's progress.

During the first seven years after 1707, for which quantitative evidence is not available, it appears that the war-induced buoyancy of the pre-Union decades was maintained. In August 1709 John Kennedy of Culzean and Hugh Hamilton, an Ayr merchant, formed a partnership to construct a saltpan and girnel at the 'maiden head of Turnberry' in south Ayrshire.[97] The ducal Hamilton family had for several years been expanding their interest in salt. In 1704 Daniel Hamilton had reported enthusiastically to the 4th Duke that he had acquired a further two pans at Thirlestane, near Bo'ness. At the same time his mother Anne, Duchess of Hamilton, was scouring the family lands on Arran in the hope of finding coal, both for shipment to Ireland as well as for saltpan use. A promising seam was found in 1708 and salters from Bo'ness were dispatched to the Auchenharvie pans at Saltcoats in Ayrshire, 'to try the coall in Arran how it made salt'.[98] The trial was evidently successful and by 1710 a saltpan had been constructed on the island, at Cock.[99] By 1713, the last year of the war with France, there were 164 coal-fired saltpans around Scotland's coasts.[100] In addition there were several peat-fired pans. Amongst those not already mentioned were the pans on Flotta in Orkney, at Portnahaven on Islay and at a number of locations on the coast of Galloway (see Map 1).

The salt officers who began to take up their stations in the latter months of 1713 may have taken some time to familiarise themselves with their

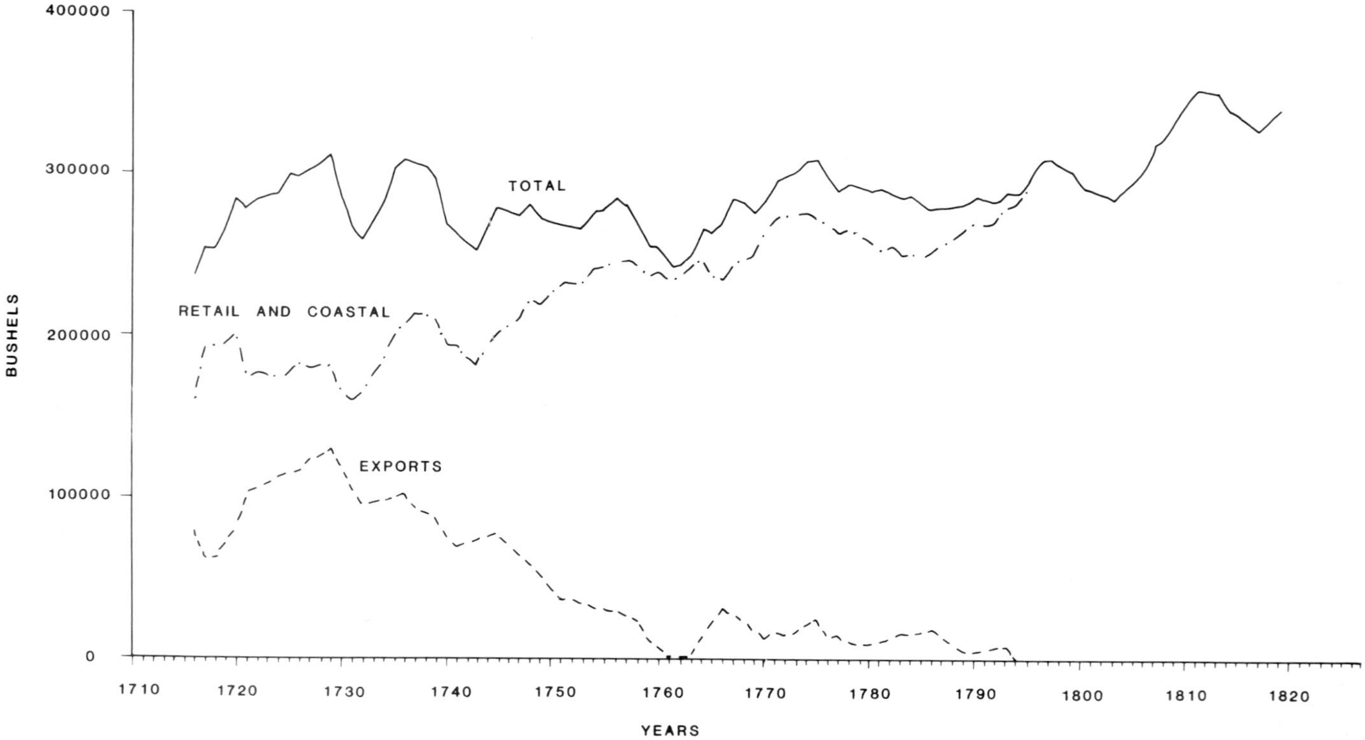

Figure 3 Scottish salt sales, 1714–1823 (5-year moving average).
Source: SRO, E 536/1-84, Salt Charge Vouchers, 1714–1798; E 551/11-18, Excise Revenue Accounts, 1798–1823.

duties and counter the resistance to Salt Office regulations displayed by some saltwork proprietors, grieves and operatives alike. After a dispute over non-payment of export duties at Methil, it was reported by a relieved Salt Office in June 1724 that the Earl of Wemyss, 'now regularly assists with their officers for the duties of all the salt he makes'.[101] The widespread failure of co-operation may account for the fact that in the year from September 1714, the first for which a complete set of salt sales vouchers is available, only some 180,000 bushels were recorded, the lowest annual figure in the whole series. Even if allowance is made for a considerable degree of exaggeration in one contemporary's estimate of a possible total of around 450,000 bushels in 1713,[102] it does represent a steep fall. Hence it seems likely that the industry received a sharp blow from the ending of hostilities between Britain and France.[103]

In the short run exports did recover, as Figure 3 shows. In the longer run however the trend was inescapably downwards and continued the pattern which had been established in the previous century. The upturn of the 1720s was probably inspired by the outbreak of plague in Marseilles and the subsequent quarantine restrictions which led eventually to salt shortages in the Baltic. As on previous occasions, commercial intelligence altered swiftly and had to be acted on decisively. Thus in January 1721 John Horsley of the York Buildings Company, then involved in the Winton saltworks at Cockenzie, was informed that they would 'do no Frats' in salt; by September demand in the Baltic was booming.[104] This led to a 65 per cent increase in exports between 1718–19 and 1723–24, on top of which followed a 31 per cent increase between 1726–27 and 1727–30, due to the recurrence of war between Britain and Spain. Thereafter competition from rock and brine-based salt shipped from Liverpool squeezed the marine salt makers of Scotland and Tyneside out of foreign markets and by the end of the 1750s no salt was recorded as having gone beyond Scottish coastal waters. Occasional single shipments were made over the succeeding decades but by 1797–8 only the recently opened saltwork at Montrose, belonging to James Dickson, appears to have sent any salt abroad.[105] If this was indeed the case the few hundred bushels involved were clearly and understandably overlooked by James Girvin in 1799 when he wrote that 'No salt manufactured in Scotland has been exported for several years back'.[106] Thus was concluded a trade which had been carried on for at least three centuries, and which in its hey-day had accounted for the bulk of the salt made in Scotland. In 1670, immediately prior to the demise of the English outlet, in the region of 200,000 bushels, or some 60 per cent of Scottish production had been sold outside Scotland.[107]

At this stage it is necessary to begin to consider the role of the Treaty of Union of 1707, and in particular the eighth Article which was concerned with the duties payable on salt, fish and meat. The immediate impact of the Union on other Scottish manufacturers varied from the devastation done to a number of Glasgow enterprises which were exposed to English competition to the rather more modest damage inflicted on the wool and linen trades.[108] In spite of claims to the contrary, salt benefited from the

Union, but not as some historians argued because the English market was opened up to the Scots, but rather because through the eighth Article the Scottish saltmasters were able to maintain their monopolistic hold over the Scottish market. Before developing this point further however the idea that the Scots gained any significant benefit from entry into the English market has to be disposed of.

After a prolonged and sometimes heated period of debate and negotiation during the last months of 1706 it was settled that when duty was imposed on salt made and sold in Scotland seven years later this would be less than one third of the rate charged in England, at 1*s*. per bushel as opposed to 3*s*. 4*d*. This fiscal inequality, it has been argued, and the absence of a countervailing duty when Scottish salt entered England, meant that 'for the next two generations the north counties of England were flooded with Scots salt which had paid less than a third of the tax paid by the Shields' salters'. While the eighth Article underpinned Scottish success in the eighteenth century, it was also alleged to have been a major cause of the demise of saltmaking on Tyneside.[109]

Yet there was an equalising duty on Scottish salt entering England.[110] For the Scots to reap the full benefits of their advantageous fiscal circumstances they would have had to conduct an illegal trade. Yet the period of their greatest potential advantage, from 1707 to 1713, when they paid no salt duties, coincided with the most successful era seen on Tyneside, when output peaked and great fortunes were made from salt.[111] An illegal cross-border trade in salt was conducted, but the scale of this can easily be exaggerated. For example, that 20,000 kitts of salmon per annum, pickled with smuggled Scottish salt, were entering England in the early 1730s, is an impressively high figure until it is calculated that as each kitt required two pounds of salt, a total of only 715 bushels or so was involved, equivalent to 0.29 per cent of the recorded legal sale of Scottish salt in 1733–34.[112] There is no evidence of shipments of salt to England from the surviving private saltwork sales books from this period. Indeed the most recent historian of the Tyneside salt industry has been inclined to discount earlier spectacular estimates of the impact of Scottish competition; the demise of saltmaking on Tyneside was principally the result of competition from Cheshire, not Scotland. The handful of English complaints about the Scots can in a large part reasonably be ascribed to prevailing anti-Scottish sentiment rather than to actual harmful competition. While some attention was paid to Scottish salt by Tyneside saltmasters in the early decades of the eighteenth century, it was the threat from the Lymington salt industry that was their major concern.[113] Where the English connection was maintained it was largely through the fishermen of Lowestoft and Yarmouth, who took on board quantities of salt from the easternmost saltworks of the Forth such as Cockenzie and Dysart on their way to the northern fishing grounds.[114]

Where the fiscal arrangements determined in the eighth Article did have a profound effect was in the Scottish domestic market. First, as before the Union, foreign salt, unless it was to be used for preserving meat and fish for export, continued to pay a heavy import tariff. The much-condemned, complex, costly and time-consuming arrangements necessary for fishermen and fish curers to comply with this aspect of the Salt Laws[115] should not be allowed to conceal the benefit which accrued to the salt industry from continued protection. Had Liverpool rock salt been allowed to enter Scotland freely there is little doubt that most domestic consumers would have used it in preference to Scottish marine salt.[116] Of equal importance in the longer run was the protection which much lower duties on Scottish salt afforded from England and the salt made from Cheshire's brine springs and rock salt deposits. Concern with the possibilities of exporting Scottish salt to England has caused historians to overlook the fact that trade with England could flow in a northern direction too. English salt had been shipped into parts of south-west Scotland from the early seventeenth century at least[117] and while there is no measurable evidence of the scale of this trade it was an issue in 1670 and had clearly grown sufficiently by the end of the century, and into the first years of the eighteenth, to cause the Scottish saltmasters considerable concern. As will be seen in Chapter Four, efforts were made in 1701 to halt the trade altogether and in 1705 an Act imposing a total prohibition on all English salt was passed by the Scottish Parliament.

Tyneside's salt industry, technologically similar to that on the Forth and making an almost identical product, peaked in the 1720s, faltered from the 1730s and declined rapidly from the middle of the century as the opening of the Sankey Brook navigation in 1757 set Cheshire's salt industry on an even steeper growth trajectory.[118] By the end of the century most of Tyneside's salt pans had either been abandoned or disappeared. Little trace of the Cumberland marine salt industry was left by the 1770s.[119] Yet as Figure 3 shows, while the Scottish industry hardly increased its output over the course of the eighteenth century, it did not decline and indeed in the early nineteenth century growth did occur and a higher level of sales was maintained until the duties were taken off. The sharp contrast in the fortunes of the Tynesiders and the Scots is largely to be accounted for by the operation of the Salt Laws. In spite of frequent changes in the level of duty, and despite appeals from a number of quarters that rock salt should be allowed into Scotland free of duty, the relative fiscal advantage in favour of the Scottish saltmasters remained. As will be seen in Chapter Four, around 1785 pressure, both public and on behalf of private interests, began to be felt, urging that the restrictions on the movement of rock salt should be lifted. Before duty the price of Liverpool salt was between 6*d.* and 8*d.* per bushel; a bushel from the pans on the Forth cost 1*s.* 3*d.* Even if freight charges were included, it was argued, if the duties were removed, Liverpool salt could undercut the Scots in the Forth itself. Scots consumers, some contemporaries reckoned, were paying over 40 per cent more for their salt than was necessary in the 1790s.[120]

As has been indicated already, quantities of salt were sold in Scotland illegally. In the 1730s for example it was alleged that 'foreign Salt was publickly Sold in all the Shops of Montrose and Arbroath'[121] while later, in the 1780s, sizeable cargoes of Liverpool salt were reported to have been unloaded at Dundee and a number of Fife ports.[122] Whatever the scale of this activity, and contemporaries' estimates differed widely, it did undoubtedly have an adverse effect on saltwork sales.[123] Nevertheless the movement of retail and coastal sales in Scotland was generally upwards from the later 1740s (see Table 2.1 and Figure 3). Indeed such was the growth in the volume of salt sold in the home market, that by the end of the eighteenth century this had completely compensated the saltmasters for their loss of markets overseas.

Table 2.1

Annual Average Sales (Recorded) of Scottish Salt, 1714–98

Period	Total sales (bushels)	Retail/coastal	Column 3 as percentage of 2
1714–20	235,400	157,025	66.71
1721–30	294,113	117,638	60.40
1731–40	287,325	193,725	67.42
1741–50	268,186	176,863	65.95
1751–60	273,666	238,700	87.22
1761–70	261,389	239,400	91.59
1771–80	294,175	269,700	91.68
1781–90	283,700	245,600	86.57
1791–98	293,744	284,441	96.83

Source: SRO, Exchequer Papers, Salt Charge Vouchers, 1713–98, E 536/1–84; note that there are gaps in the series, 1715–6, 1719–21, 1730–2, 1740–2, 1743–4, 1765–6, 1777–8, 1779–80

The table shows that by the final decade of the century annual sales of Scottish-made salt in the home market amounted to an average 97 per cent of the total. As far as total sales are concerned perhaps the most notable feature of the figures is the absence of any significant trend either upwards or downwards after 1720. The fortunes of the salt trade fluctuated, with the worst decades being the 1710s and 1760s. The best were the 1720s, 1770s and 1790s, with average annual sales in the latter period being marginally below those of the 1720s. Nevertheless in 1796–7 over 320,600 bushels were sold, more than had been recorded in any previous year, heralding the beginning of a generally buoyant period which lasted until 1825.

The scenes of decay—the ruined and deserted saltpans—which were painted by some of the contributors to the *Old Statistical Account* in the early 1790s, were neither typical of the decade, nor indicative of long-term decline, but represented instead the end of a trough which was to precede a final flurry of saltmaking activity which lasted almost thirty years. The changing situation is captured best in the *Account* for Kirkcaldy. Salt was 'once' manufactured in the burgh, it was reported, but had virtually ceased since the local supply of coal had been exhausted. A 'late advance' in the price of salt however, caused by the outbreak of war with France, had been sufficient to rekindle interest in the industry locally. How substantial the upturn was after 1793 can be seen at Pinkie, where average annual sales in the five years up to November 1793 had been 15,544 bushels. The equivalent figure for the following five years was 24,084, an increase of some 54 per cent.[124]

Annual average sales in the second decade of the nineteenth century were not much below 340,000 bushels, that is around 15 per cent higher than the levels achieved in the 1790s. As far as the eighteenth century as a whole is concerned though, the industry was not in decline, nor had it experienced great prosperity. Rather, it had undergone a process of transition, shedding its former dependence upon foreign markets and the fortuitous effects of European wars, and continuing to build upon the domestic market. At the century's end, an old ally, wartime dislocation, came to its aid again, just as it had a century earlier. This time however, it was only in Scotland itself that sales of Scottish salt could be achieved.

On the demand-side, the broad movements in retail and coastal sales appear to have been closely linked with the patterns of population change which have been identified in recent research into Scottish demographic history. A period of expansion early in the eighteenth century was followed by one of just over twenty years (from 1720) when little growth occurred. Thereafter Scotland's population probably grew continuously, if unsteadily, until the early decades of the nineteenth century when the rate of growth increased sharply.[125] The boom in Scottish salt sales which occurred after 1806—there was a 29 per cent increase between the months of July 1805–6 and 1812–3—reflected not only the rising rate of population growth but also to some extent the impact of the Napoleonic wars, although by this time Bay salt had largely been supplanted by Liverpool suppliers. Industrial demand for salt was rising too, particularly from bleachers who used it as a raw material in the manufacture of chlorine. Soda, from salt, was used by hard soap makers. Glauber and Epsom salts too were made from marine salt.[126] High duties however were to deter chemical manufacturers from using salt on a large scale,[127] and after they were removed it was Liverpool rock rather than Scottish marine which they preferred to use. In 1813–4 duty was paid on 4,179 bushels of salt used to make Glauber and Epsom salts in Scotland, which was equivalent to just over one per cent of all Scottish salt sold; the 622 tons of salt used to make 'muriatic acid' in 1821–2 amounted to around 6.5 per cent of the total.[128] Whatever the respective shares of its component parts though, demand was certainly intense. Between 1809 and 1817 as many as sixteen new saltworks may have been erected on the Forth alone (although cut-throat competition caused almost as many to close) while in 1814 works were established by James Butchart at Arbroath. Also in 1814, for the third time in two centuries, saltpanning was recommenced at Brora in Sutherland;[129] in the same year the site of the 'ruinous' saltpans belonging to the earls of Eglinton at Ardrossan was re-let and three new pans built. Also in Ayrshire, a short-lived set of pans was constructed at Irvine.[130]

The hypothesis that movements in retail and coastal sales were causally connected with demographic change rests upon the quite reasonable assumption that per capita salt consumption remained roughly constant. There were periods however when this was not so. The sharp fall in home sales which occurred at the end of the 1730s, revealed in Figure 3, was almost certainly the result of additional hardships experienced by the population following the poor harvests of 1739 and 1740. Following a pattern discernible in other consumption goods such as beer and whisky, salt sales did not recover their 1737-8 levels until 1745-6.[131] Statistical difficulties rather than historical influences probably account for the sharp dip which marked the very late 1720s and the first years of the 1730s.[132] As will be seen in Chapter Four it was the operation of the Salt Association rather than raised mortality levels, subsistence crisis or market forces, that produced the steady decline in sales which began in the early 1770s and lasted until the middle of the following decade.

It is important however to recognise that behind this account, which has swept broadly over time and space, are concealed significant regional and local variations from the general trends. Market circumstances were far from uniform in the post-Union Scottish salt industry. The significance of the industry's transition from fairly heavy dependence on exports, to total dependence upon its home market, was far greater in Fife than anywhere else in Scotland. Early in the eighteenth century, some 83 per cent of the country's salt exports were being sent from Fife saltworks, principally those at Kirkcaldy, Kirkcaldy Links, Dysart, Methil and Wemyss.[133] The last two relied more heavily than most on their customers overseas, who accounted for 64 per cent of their sales between 1716 and 1719. Most of their salt went to places such as Bremen, Campvere, Danzig and Rotterdam, which in the seventeenth century had been in receipt of salt from a much wider range of Scottish salt producing precincts. During the course of the eighteenth century, as exporting opportunities dried up, and in the absence of sufficiently strong demand locally, the direction of salt shipments from Fife shifted further towards Scotland's north east coast. Although coastwise links had been established with the north beforehand, shipments to places such as Banff and Wick were far less significant than those to Edinburgh and especially Leith, from whence much of their salt had been shipped overseas. By the 1780s, 80 per cent of the salt shipped coastwise from Wemyss and Methil was being sent to ports north of Fife, such as Dundee, Perth, Montrose, Aberdeen and Nairn.

In sharp contrast, virtually no salt was exported from the five works which were in existence in Ayrshire in the early eighteenth century.[134] Similarly, while there were notable exceptions on the south bank of the Forth, as at Cockenzie and Thirlestane for instance, many works had significantly reduced their dependence on overseas sales, and were exploiting instead the growing urban markets of Edinburgh and the Lothians. Not surprisingly, from those works which were closest to the capital, as in the cases of Edmonstone, Westpans, Cuttle and others in the Prestonpans customs precinct, only occasionally was salt shipped either

6 Salt being removed from the pans, Prestonpans 1958. Usually this was then put in 'drabs', in which the salt was partly dried, before being stored in the larger 'girnel' (sometimes 'girnal') from where it was drawn for sale. Photograph by S. F. Sanderson, *School of Scottish Studies, University of Edinburgh.*

coastwise or overseas, but instead was dispatched to customers by road, either in carts or on the backs of horses and women.

The prop which had sustained the industry since the 1660s, the protected home market, was kicked away in 1825 when the last salt duties were lifted. The danger which repeal posed for the Scots was put succinctly in the *Scottish Monthly Magazine* in 1836: 'The great advantage which Cheshire possesses in the brine springs over other parts of the kingdom is strongly indicated by this Fact, that when...the duty on Cheshire salt was 15*s*. a bushel, that obtained from sea water in Scotland could stand a duty of no more than 6*s*. while, since the abolition of duties altogether, the competition of Cheshire has almost entirely extinguished the manufacture of Scotch salt'.

Indeed, within a few years many works had closed, while others were much reduced. Thus the venerable works at Wemyss, which in its hey-day had boasted eight pans, had only one left when the *New Statistical Account* was written. Over the river Forth at Tranent the number of pans had been halved to six. The sharp fall in output was mirrored in reduced profits. At Dysart colliery and saltworks, also in Fife, the profit on salt fell from a peak of just over £692 in the year to May 1825 (the immediate effect of repeal was a short-lived increase in sales prior to the inrush of salt from Liverpool) to around £91 in the following twelve months. By the 1860s the works had closed. Saltwork income at Fordell fell less steeply, from an annual average of some £443 between 1823 and 1825 to an average of £304 ten years later.[135] In no single year after 1825 though, did returns from the pans ever come near to the £951 which they produced in 1824.

Where the industry survived it usually did so by importing Liverpool rock salt and refining it. Dissolved in sea water it greatly strengthened the brine, reduced evaporating costs and made an 'excellent' salt for which there was still considerable domestic demand, especially for cooking.[136] There were even periods of acute shortage, especially prior to the completion of the Anglo-Scottish rail links, in mid-century. In November 1846, for instance, the salt agent at Fordell had to inform James Miller, a Perth merchant, that 'the demand for salt is so great I am doubtful whether I will be able to load a vessel full up'.[137] While bleachers and other chemical manufacturers in Scotland usually but not always satisfied their salt requirements by bringing salt directly from Liverpool, there was still sufficient life in the domestic market, especially for culinary salt, to justify some new investment in rebuilding and improving pans and panhouses, including for example the purchase of steam engines to pump sea water into the works' reservoirs. Even as late as the 1860s the saltworks at Joppa were evidently improved and enlarged.[138]

The trend was generally downwards however and from before the mid-nineteenth century it became increasingly difficult for owners to let their pans. Indicative of the way the world was turning was the notice of sale of Pinkie saltworks in 1855. Considered as important as the saltmaking equipment, workers housing and ancillary material, was the fact that the site had 'a frontage of 360 feet to the high road to Edinburgh...well suited for a

7 Weighing and packing salt, Prestonpans 1958. The use of cardboard boxes was a twentieth century development. The Scottish Salt Company however was formed in 1889. The other noteworthy link with the industry's earlier history was the fact that the salt being weighed was coarse cooking salt, underlining its long-standing domestic utility. Photograph by S. F. Sanderson, *School of Scottish Studies, University of Edinburgh.*

manufactory, or Iron ship building yard....'[139] The only works still making salt from sea water alone was at Saltcoats in Ayrshire, where the proximity of the dairy industry and makers of Dunlop cheese ensured the lingering survival of 'Ayrshire marine' manufacture until the late 1870s. In 1889 though, the saltworks at Bo'ness closed down. In the same year, in a defensive measure designed to eliminate competition between most of the remaining east coast works as well as to reduce working expenses by 'supplying each district from the most convenient Salt Works' the Scottish Salt Company was formed. Partly modelled on the English Salt Union, the Scottish Salt Company had an initial declared capital value of £65,000, although only £22,500 was called up at first. The average stake of each of the 12 shareholders (most of whom were individual proprietors of small saltworks) was £1,875. Some years later the shareholders voted to reduce the Company's capital to £47,485 and by 1914 its total capital value was estimated to be only £9,918. A number of its component parts did struggle on through the first half of the twentieth century though, and it was not until June 1964 that the Scottish Salt Company was finally wound up. In name at least, it had survived the last going Scottish saltwork by five years.[140]

DOMESTIC SALT: PATTERNS OF SUPPLY AND MEANS OF DISTRIBUTION

Having established the importance of domestic salt sales to salt work proprietors in the eighteenth century, and the key role which Scottish-made salt played in satisfying the salt needs of the Scots population, it is now necessary to describe briefly the mechanisms by which producer and consumer were linked. The main features of the operation of the domestic salt market in Scotland can be identified fairly readily. In transport terms, the means by which consumers were provided with salt was determined largely by their geographical location and proximity to salt pans. Prior to the railway age there were only two ways in which salt could be transported from the pans to markets: by coastal shipment or overland. Over the course of the eighteenth century, the proportions of salt sold from the pans which went by these means did not greatly differ, although there was some shifting of their relative importance. In the period 1716–19 for instance, the proportion of Scottish salt sales which are recorded as having gone coastwise averaged 30 per cent per annum, while landsales accounted for 43 per cent of the total.[141] By the 1770s coastal sales accounted for a slightly higher proportion of total sales than landsales, but by the 1790s landsales again took a bigger share, 51 per cent as opposed to 47 per cent which went coastwise.

That such a sizeable proportion of Scottish salt was being transported by land carriage in the eighteenth century can largely be accounted for by the proximity of Edinburgh and the relatively densely populated Lothians to the heartland of Scottish salt production, the Forth.[142] 'Cageors and other poor people', had long taken salt from the pans, either for their own use, or for resale. Best known perhaps were the 'saut wives' of the parishes of Duddingston and Inveresk, who carried salt purchased from the pans at Joppa and Pinkie on their backs in baskets and creels into Edinburgh.[143] This trade, it appears, was almost solely the preserve of females who were not only physically strong, but also capable of driving a hard business bargain. Around the turn of the nineteenth century, it was claimed, they were buying salt from the pans for around 1*s.* 1*d.* for a peck of seven pounds weight, and reselling it in quarter peck measures for 6*d.*[144] There may however have been an element of exaggeration here, for as late in 1793, when salt prices had temporarily risen to an exceptionally high level, it was considered little short of scandalous that retailers or shopkeepers in Edinburgh had raised the selling price of a full peck of salt from 9*d.* to 1*s.* 3*d.*[145]

Small-scale salt carrying and hawking however was by no means confined to Edinburgh and the towns surrounding it. Salt from Culross and other saltworks nearby was taken by salt carriers such as John Marshall of Menstrie deep into Perthshire, as far north as Callandar and Doune. Salt from the same pans was also carried into Stirling, north Lanarkshire and east Dunbartonshire.[146] There were however geographical limits to the small-scale carrying trade from the Forth. The Glasgow market appears to have been increasingly dominated by merchants who purchased salt from the east coast pans in bulk, as opposed to the handful of bushels with which the individual carriers dealt. Furthermore, the further west they went, the greater was the likelihood that they would be faced with competition from salt which had been smuggled in from Ireland. This was a major source of supply in Galloway, and appears to have made a fairly significant contribution in Ayrshire.[147] Glasgow too was freeuently served with salt from this direction. Occasionally it could even be found in Perthshire, much to the consternation of regular carriers and the salt officers, who reported in 1796 that the carriers of Irish salt, 'went in number together' and were 'Dangerous to Meddle with'.[148] They did not however venture into Berwickshire, Peebleshire and the other Border counties, which were supplied overland from pans on the south side of the Forth, principally from Cockenzie and Prestonpans.[149] Supplied neither by legal traders nor smugglers during the winter months, were the inhabitants of the inland parts of counties such as Argyll and Ross. As a consequence they either had to pay exorbitant prices for what could be obtained or do without.

For people who lived nearer to a saltwork, it was probably cheaper to buy direct, although the prices charged at the pans for 'Country sales' were usually at least ten per cent above those charged to either coastal shippers or salt exporters, who tended to buy in greater bulk. Hence it was on local 'inland' sales that saltmasters reckoned to make their highest profits per bushel.

'Cadgers' too paid a slightly higher price, although this was less than individual consumers were charged. At the Earl of Eglinton's pans in 1757 for example, 'cadgers' were charged 6*s.* per boll, while 'Country' purchasers paid 6*s*. 6*d*.[150] As partial compensation salt carriers who had remained loyal to a particular saltwork for a period of time might be granted a rebate or bonus, as happened in the case of 'Crighton a Salt Cadger', who was given an 'abatement' of 5*s*. 6*d*. by Dysart saltworks in June 1817 'for his good Custom'.[151] The quantities purchased directly by consumers from the pans were invariably small. Thus of the 56 individual sales 'to the contre' from Dysart between June and August 1717, 19 were of less than a bushel, while on only four occasions were as many as two bushels sold.[152] A similar situation can be seen from the accounts of Turnberry saltworks in south Ayrshire, although in this rural district customers appear to have been forced to travel several miles from inland villages such as Cumnock and Maybole in order to purchase relatively small quantities of salt.

Salt need however cut through distinctions of class or status, with Sir John Cathcart, Sir Thomas Kennedy and John Goudie, minister of Maybole, jostling in the sales accounts alongside the names of tenant farmers, tradesmen and common people.[153] The only difference was that the higher up the social scale customers were, the greater were the quantities they purchased likely to be. They were better able to afford to buy several bushels or a boll of salt at once, and needed more salt to preserve and season the more substantial food requirements of their larger households. The Earls of Eglinton expected their saltwork lessees to provide the occasional boll of salt for the use of their farm horses.[154]

The further people were from a set of pans however, the greater was the likelihood that their salt would be brought at least part of its journey by sea. Thus the inhabitants of Scotland's east coast towns and settlements from the rivers Tay to the Spey, and sometimes even further north, depended almost wholly on coastal shipments. While cargoes of salt could be shipped on their own, it was more usual for another cargo, often coal, to be carried along with it. In December 1800 for instance, John Chalmers of Banff asked John Rose, the salt factor at Wemyss, to send him 1,500 bushels of salt, 'and what Great Coals may be required to load up the vessel'.[155] The principal places to where salt cargoes were shipped, before being broken down for local urban sales and distribution inland, were Dundee, Perth, Montrose, Stonehaven, Aberdeen, Peterhead, Fraserburgh, Banff, Speymouth, Fortrose and Inverness.

There were times though when the north east's dependence upon the Forth was at least partially relieved. As has already been pointed out, early in the eighteenth century there was an active market in some east coast ports in illicitly entered Bay salt, diverted from its legal use as a preservative of fish for export. At the turn of the seventeenth century, the inhabitants of Sutherland and its 'neighbouring provinces' were supplied with salt from the pans erected at Brora by Jane, Countess of Sutherland in 1598.[156] The re-establishment of salt making there in 1814 again reduced the area's dependence upon imports from the Forth. Indeed it was the construction in 1794 and 1795 of a string of saltworks which stretched from Dundee to Portsoy, which probably accounts for the fall in coastal sales of salt which occurred during the 1790s, as Fife saltmasters in particular found themselves facing growing competition from local producers in their critically important market in the north east. In the second half of the 1780s for instance, 80 per cent of the coastal shipments leaving the saltworks of Methil and Wemyss, went to the modern local authority regions of Tayside and Grampian.[157] By building pans at Usan, David Scott of Dunninald hoped to supply a third part of the salt needs of the county of Angus. The four pans at Brora, it was claimed, ultimately satisfied a good part of the demand for salt on the Moray Firth, at two-thirds of the previous price.[158]

The high price and poor quality of Scottish salt on its arrival in the north east had been a matter of considerable resentment from the 1660s, when it had become the only salt which could be used legally for domestic purposes. It 'cannot be serviceable either for towns or countrey for befor it could be transportit to remote parts of the iland it sould turne all into water', protested the Town Council of Aberdeen in July 1665.[159] Leaking hulls, rotting decks, and the tendency of Scottish salt to liquify, made its transportation by sea a hazardous business. It was largely for this reason that salt purchasers in the north east were able to obtain an allowance from panmasters for wastage en route. This was more often a customary rather than a formal arrangement: concerned that a sea voyage during winter was more likely to lead to salt wastage than at other seasons, in the aforementioned order for salt from Wemyss, John Chalmers of Banff could only urge his supplier 'to desire your Servts to give good Weights'.

Shipments sent north from pans on the Forth were rarely below five or six hundred bushels, ordered by merchants located in the north eastern ports. They then redistributed the salt both locally, to retailers, and for transportation inland. Merchants who did regular business with the same works could frequently obtain three months credit, as in the case of Provost Simpson of Perth, who in the 1770s was purchasing about a quarter of all the salt made at Methil and Wemyss. He also appear to have been the major supplier of domestic salt in and around Perth.[160] 'Transient customers' however, usually paid for their cargoes in cash.[161] In Ayrshire too, the salt market in the county town appears to have been concentrated in few hands, notably those of George Black, who did temporary damage to the fortunes of the Craigie works by importing salt from Saltcoats when the Craigie lessees refused to drop their price in 1777.[162] On the other hand the absence of other large towns, with the exception of Kilmarnock, and the close proximity of saltpans, caused the salt trade to be dispersed more widely than was the case in the north east. Not only were 'Cadgers' frequent visitors to the pans at Saltcoats, but it was also not uncommon at Turnberry for very small orders for salt to come from what appears to have been very modest purchasing collectives. Thus in September 1757 one and a half bushels of salt, worth 2*s*. 6*d*., was sold to 'David McMester in Arran & others'.[163] While most of the salt made in Ayrshire was sold within the county, some was shipped further afield, notably to merchants in Dumbarton, Greenock and Galloway.[164]

Although most of the salt needs of the inhabitants of Galloway appear to have been met through a combination of the activities of smugglers and what was made at the many small pans which operated in the area, after the middle of the eighteenth century some large shipments of Forth salt found their way into the region. Just why this happened is not altogether clear, although population growth and the disappearance of most of the region's saltworks by mid-century almost certainly played a part.[165] In the late eighteenth century too the trade in salted meat was growing, although there was considerable reluctance amongst curers to use Scottish salt as a preservative. The 'peculiar flavour' it gave to bacon and hams seems likely to have been attractive to only a minority of consumers.[166]

Whether or not salt from Ireland and the Isle of Man became more difficult to obtain cannot be easily ascertained, although that considerable anxiety about the extent of the illicit trade was still being expressed in the 1790s suggests that there remained a great deal of life in this activity. Indeed if one contemporary estimate is to be believed, it was only in exceptional circumstances that sellers of Scottish salt could hope to compete with the smugglers. Shipped at Larne for 1*s*. 1*d*., per bushel, and selling in Scotland at 2*s*. 6*d*., illegally imported refined Liverpool rock salt was clearly in an extremely advantageous position at a time when the same quantity of Scottish salt, including duty, cost 3*s*. 2*d*.[167] The profit margin on a bushel of smuggled salt was reckoned to be 1*s*. 5*d*. in the 1790s, although this had to cover the risks of sinking somewhere in the Irish Sea, and of being caught smuggling. The basic cost of making a bushel of sea salt around this time was in the region of 1*s*. 3*d*. or 1*s*. 4*d*. At times it could even be higher.

What is certain however is that by the mid-1770s many saltmasters on the Forth, acting in unison under the flag of the Salt Association, were anxious to reduce the quantity of salt available for sale on the east coast, in the hope of raising its price. A subsidy which was available to their members who sold salt well beyond the natural markets of the Forth was a sufficient inducement for some saltwork proprietors to search for outlets in the west, while its reduced price evidently made it attractice to salt merchants in Dumfries. The activities of the Salt Association deserve to be examined in greater detail however, and form part of the subject matter of Chapter Four.

3

SALT, COAL AND CAPITAL

THE SALT-COAL RELATIONSHIP

Traditionally, the salt industry has been seen very much as the junior partner of the coal industry. An 'offshoot' or appendage of the latter, it has even been argued that the salt industry's output patterns were a barometer of the fortunes of coal, growing during periods of expansion; doing less well when coal mining was in the doldrums.[1] Yet as will have become apparent from the previous two chapters, on the supply side, many factors, amongst which the availability of coal was only one, determined how much salt could be put out from the pans, while demand was influenced by war, tariffs, demography, and the effective strength of competition from other salt producers. The notion that Scottish salt sales simply followed movements in coal output does not stand up to scrutiny: peaks in exports of salt from Scotland in the 1480s and 1530s for example were not shared by coal,[2] while the trebling of Scottish coal production in the eighteenth century did not produce a similar increase in salt output. It can also reasonably be argued that historians have uncritically read back into previous centuries what was the situation in the late 1700s and early nineteenth century, when, compared to the value of the output of the coal mining industry, as well as the capital and labour resources devoted to it, salt manufacturing was of only minor significance, in a few coastal towns. Only at this stage did salt unquestionably become the junior partner.

The purpose of this chapter, then, is to re-examine the relationship between coal and salt. In so doing it will become clear that in the early modern period, salt played a far more important role in the economics of coal mining and the landed estate than has previously been allowed. In turn this further accounts for the intense interest which the subject of salt aroused at the time of the Union of 1707, mentioned in the Introduction. Indeed only with an appreciation of the proper place of salt in the fortunes of a number of Scottish landed estates and joint colliery-saltwork enterprises can the political activities of the saltmasters, the subject of Chapter Four, be fully understood.

It cannot be denied that the coal and salt industries were intimately linked. 'A tour of the mining districts', Professor Duckham has rightly written of the eighteenth century, 'would have revealed few of the leading coal masters with no interest in salt.'[3] With all but the smaller rural pans being dependent upon coal fuel, it made a great deal of business sense for coalmasters to own salt pans. Salt sales contributed to the income of joint coal-salt enterprises. They also provided an almost certain market for the otherwise largely unsaleable 'small' coal, much of which was put out from Scottish pits. Although liming of agricultural land had been done at least as early as the sixteenth century,[4] it was not until the later eighteenth century that this activity became widespread and created intense demand for coal to heat the numerous lime kilns which were then being built.[5] Similarly, industry, which also made use of small coal, for brewing, glassmaking and iron manufacturing for example, only began to make noticeable demands on Scottish collieries from the 1760s.[6]

It is easy, however, to underestimate the extent to which Scottish coastal collieries depended upon sales of small coal to the salt pans, as well as the income from the salt itself, prior to this period. At times they provided not merely a 'useful' outlet, allowing coalmasters to spread their risks and utilise all of a colliery's productive capacity, but an outlet without which they would not have been able to carry on their other coal mining operations. 'The coal mines indeed were at one period largely worked to keep the pans boiling', wrote an historian intimately acquainted with the Forth valley coal industry.[7] Just what period Cadell was referring to can only be speculated upon before *c.*1670. Only after that date have sufficient numbers of colliery-saltwork account books survived to make possible a quantitative assessment of his judgement. It seems reasonable however to suppose that before the mid-sixteenth century, when little coal was exported from Scotland, and little burned in domestic fires, the salt pans may have been the main outlet for Scottish coal.

The great expansion in exports of coal from Scotland, which lasted for just over a century from the later 1500s, is likely to have reduced the collieries' dependence upon the saltworks. However, with deeper mines being sunk, necessitating the use of relatively expensive pumping apparatus, the additional income they provided was still judged by many to be vital. Without 'the benefite of the salt', it was claimed in a petition to the Privy Council in 1631, 'these sumptious water workes and mynes...cannot be upholden'. At times of crisis, when trade was interrupted, qualitative evidence suggests that saltwork demand for coal was particularly important. In correspondence sent to the Marquess of Tweeddale on behalf of eleven of the country's leading coal and salt proprietors in 1665, for example, it was admitted that a recently imposed tax on imported salt, had, by protecting the Scottish salt industry, allowed them to maintain their collieries as well as their saltworks, 'which without that remedy had been...ruined'.[8] This was at a time when coal exporters were still experiencing the damaging consequences to their trade of the Civil War.[9]

While the coal trade recovered in the 1670s and 1680s, by the end of the century, many Forthside collieries were in considerable difficulties. On the one hand, competition from pits on Tyneside and at Liege, better fuel husbandry in the Netherlands, and the imposition of new excise duties and tariffs, did much to undermine the Scots' exporting efforts.[10] At home, decades of intensive mining of the most accessible shoreside seams, and problems about the desirability of the type of coal being brought up—it burned brighter but less economically than that mined on Tyneside[11]—added to the industry's difficulties. On top of all this came the crisis of the 1690s, the so-called 'Seven Ill Years'. The devastating effect of the unusually bad weather was not confined to the agrarian sector. The 'great floods of rain' in 1695 and 1696 raised mining costs to prohibitive levels, as colliery proprietors were forced to employ men and horses around the clock to keep water levels in the pits down. High grain costs, and occasionally the inability of factors and grieves to obtain oats at any price, compounded the problem. Sometimes, the additional water simply could not be managed, and pits had to be temporarily abandoned, as occurred at Kinneil for six months in 1698.[12] The advent of a new century brought no relief: 'excessive rains' continued to fall, and in February 1702 it was reported that the 'whole Coals on both sides of the water [the river Forth]', were in an 'ill condition'.[13]

With markets for coal disappearing and production difficulties mounting, it seems scarcely surprising that there was an almost universal conviction amongst the proprietors of joint colliery and saltwork enterprises at the turn of the eighteenth century that 'the whole Advantage of Scots Coalmasters Depends on their Salt'.[14] As was seen in Chapter Two, the market for salt was then exceptionally buoyant. Thus in 1707, one interested party could claim that most collieries depended 'intirely' on the saltworks. If they failed, the coal industry would fail too, 'for if we have no use for our Pan-wood, then the Profits arising from the Work will not support the charge.'[15] The extent to which this was so can be seen at several locations. On the ducal Hamilton estate of Kinneil, it is notable that when flooding left the factor with no alternative but to abandon their then principal pit in 1702, great care was taken to ensure that a supply of small coal for the pans was maintained.[16] At the neighbouring Grange colliery, a striking shift in the direction of coal output underlines the notion of a late seventeenth and early eighteenth century crisis in Scottish coal mining, and gives force to the suggestion that dependence upon salt was growing. In 1695 over half of the colliery's output was 'sea coal',and shipped. By 1701 the pans were consuming 87 per cent of the coal put out, while sea sales had fallen to an insignificant 22 loads.[17]

An extremely interesting example comes from Tulliallan, one of the largest coal and salt concerns on the Forth.[18] In 1679, some 93 per cent of the colliery's output went to the salt pans.[19] The date is important, as it suggests that well before the critical period of the 1690s, consumption of coal by the salt pans may have been substantial. It has been estimated that exports of Scottish coal in the seventeenth century are unlikely to have been greater than 30,000 tons per annum, and often much less.[20] Although Professor Nef's oft-quoted suggestion that 150,000 tons of small coal were used by the saltworks in the late seventeenth century is greatly exaggerated,[21] even a modified estimate of coal consumption, of 50,000 tons per annum around 1670,[22] is indicative of how significant a coal user the salt industry was.

If because of a lack of data there must remain some uncertainty about the relationship between coal and salt at the level of the individual enterprise much before 1700 (although in general it seems that coal was heavily dependent upon salt), there is much less reason for doubt over the following century and a half. Domestic demand for coal grew little, if at all, until around 1760.[23] Coal exports from the Forth probably failed to recover to their best seventeenth century levels throughout the 1700s, and indeed at the end of the century may have been below 12,000 tons.[24] Even rapid rates of export growth from Ayrshire to Ireland should not conceal the relatively low tonnage which was being shipped from the county's three main harbours by 1800. More spectacular advance occurred only in the early-mid nineteenth century.[25]

In spite of the optimism with which some historians have viewed Scottish colliery profitability in the eighteenth century,[26] these and other considerations provide substantial justification for an alternative, pessimistic, conclusion. Robert Bald, for example, the celebrated coal 'viewer', who was intimately acquainted with the Scottish coal industry, reported that since the introduction of the steam engine, 'few or no fortunes' had been made from coal mining. Even without the capital and running costs of Newcomen steam engines, colliery owners and lessees found it expensive to keep their necessarily deeper pits dry. William Adam, for example, the renowned Scottish architect, had combined his work on designing and building Yester House for the Marquess of Tweeddale with mining coal on the Yester estates at Pinkie. He was, he wrote in June 1743, 'at an intollerable expence for Drawing water by Horses which runs away with the profits and is frequently more'. Two decades later, in 1764, his son John was also bemoaning the family's losses over Pinkie coal, but, he wrote, despite, indeed partly because of, the past expense and the future expenditure required, 'there is no looking back.'[27]

Scottish colliery labour costs were also high. From mid-century, wage levels appear to have risen above those paid in England.[28] The strength of organised labour too may have acted as a damper on coalmaster enterprise.[29] The quality of Scottish coal continued to be a problem, and the proportion of small coal brought up was too great. Indicative perhaps of poor market prospects, with a few notable exceptions, Scottish coalmasters appear to have been slow to invest in Newcomen steam pumping engines. It is little wonder that in March 1777, when writing to Samuel Mitchelson WS, about his struggling colliery at Culross, Archibald Cochrane, 9th Earl of Dundonald, prefaced his report with the sentence: 'You will naturally say that those that meddle with Coal works, should be possessed of the Philosophers Stone.'[30] For several reasons then, until the general strengthening of domestic and industrial demand which occurred after 1760, many coastal collieries which did not have easy access to the substantial and growing markets of Edinburgh and Glasgow in particular, depended for their survival upon their saltwork operations.

There were of course exceptions. On the upper reaches of the Forth, where, owing to the greater presence of fresh water, salt making was both difficult and expensive, the industry was conducted at places such as Clackmannan on a diminishing scale by the early eighteenth century. Thus, at the Earl of Mar's Alloa pits, strenuous efforts were being made to increase 'great' coal output and reduce the proportion of 'chow' coals brought to the surface.[31] Salt production on the Mar estate appears to have all but disappeared before the end of the seventeenth century.[32] At Limekilns, where at least one of the discarded Alloa pans was rebuilt, the Halketts of Pitfirrane were also far more interested in coal than salt. Largely, this was because while their coal was described as 'only fit for the Smiths forge' (which was ideal for salt pan use), they did not have to turn to large scale salt manufacturing. Under the terms of the VIth Article of Union, which protected certain private rights and exemptions granted by the Scottish parliament prior to 1707, Pitfirrane coal was relieved of English taxes on coal exports and could therefore be sent abroad at a competitive price.[33]

Other owners of coastal collieries, along with their lessees, found themselves in very different circumstances. One of the most striking examples comes from Bogie coal and saltworks, at Kirkcaldy. There, as Table 3.1 shows, in the three accounting years from 1719–20 until 1721–2, the large clear profits from salt sales more than compensated for the losses made on coal mining operations.

Table 3.1

Profit and loss accounts, Wemyss of Bogie Coal and Saltworks, 1719–22 (October–October)
(+ = profit; - = loss. Figures in Scots money values)

	1719–20	*1720–1*	*1721–2*
Coal	*-1,267.5s.*	*-1,446.12s.*	*-2,142.12s.*
Salt	*+2,970.7s.*	*+5,598.9s.*	*+4,793.5s.*

Source: NLS, Wemyss MSS, MS 5723, Tutors Sederunt Book, 1719–33.

Similar, if less striking, examples come from other joint coal and salt enterprises in Fife in the middle decades of the eighteenth century.[34] Even in Ayrshire, where coal sales to Ireland were growing, the county's largest and most technologically advanced colliery, at Stevenston, depended entirely on salt sales to outweigh the losses made on coal between 1749 and 1761.[35]

Less than a decade later, Dr John Roebuck, then the lessee of Bo'ness colliery and saltworks, found that even from a new pit sunk to the 'seven foot coal', 'a prodigious quantity of Panwood, in proportion to the other species of Coals', was being put out. Of two alternatives, either to alter the colliers' underground working practices and method of cutting the coal, or to construct more saltpans, the latter was judged to be the sounder course.[36] The success of the joint venture, he argued, depended upon his ability to make 'the Saltwork a profitable branch of Business at all times'. Also at Bo'ness, the partners of the Grange Coal Company resolved in 1779 that it was 'of the greatest consequence to the Company to keep the pans...fully

8 'The Bucket Pats', Kirkcaldy. This rock bathing pool at Kirkcaldy, photographed at the turn of the twentieth century, had, in earlier times, a far more practical function, as a 'bucket pot' for one of the local saltpans. Sea water ran into the pond or reservoir as the tide flowed, through the cut which can be seen to the rear. The water could then be drawn into the saltpan as and when it was required when the tide was out. For further details see Chapter 1. From the collection of Eric Simpson, Dalgety Bay.

and ably going', and that if the main coal did not produce enough panwood, 'an opening should be made in the Foul Coal in order to get a full quantity raised'.[37] Both Roebuck and the Cadells of Grange were members of the Salt Association. Significantly, one of the principal reasons for the Association's formation, was the conviction of its members, that any change in the salt duties, the operation of which had protected Scottish salt in one way or another since the 1660s, would ruin the Forth salt trade and be 'highly prejudicial to their Interests as Proprietors of Coal'. As a recent study of Halbeath colliery and saltworks in the period 1785–91 has demonstrated, even at this late stage in the eighteenth century, the saltworks 'were not merely a profitable adjunct to the colliery...they constituted a vital part of the economy of the whole enterprise.'[38]

By the end of the century however, such instances were rather less common, although the fortunes of some collieries did continue to be integrally bound up with those of their saltworks. It was reported of Dysart colliery, in 1817, that without the consumption of the small coal by the salt pans, 'the Dysart Coal cannot be wrought on account of its spontaneous combustion if left below ground'. The costs which would be incurred in bringing it up and simply leaving it unsold could not be borne.[39] Intense competition from imported Newcastle coal in many Scottish towns made it difficult for the factor at the reopened colliery on the Sutherland estate at Brora to find markets for his coal, and thus in 1814 it was anticipated that salt, and to a lesser extent lime, would produce the integrated works' greatest profits.[40]

Generally though, as the output trajectory of coal steepened, coastal collieries freed themselves from their dependence upon salt. Increasingly, income from coal sales overshadowed that from salt. On the estate of the Hendersons of Fordel, near Inverkeithing, for example, in 1791, the profit from the colliery, £3,383.10*s*, was almost nine times greater than that returned from the saltwork.[41] Even on the St Clair's Dysart estate, in spite of the strategic importance of maintaining a saltwork, colliery receipts averaged some £5,350 in each of the two years from February 1818, while saltwork receipts averaged £1,084.[42] With the removal of protection for Scottish salt in 1825, and as coal output expanded to feed the fires, forges and furnaces of industrial Victorian Scotland, the gap between colliery and saltwork incomes widened further still.

The extent to which collieries continued to be dependent upon their salt making operations and sales of small coal to salt pans in the later eighteenth century, is however confused in the account books of joint enterprises by the failure of individual coalmasters to resolve the vexed question of how much they should charge for the small coal used at the pans. Clearly, as some contemporaries became increasingly aware, if the market value of the fuel was not charged to the pans, profits apparently made from salt actually belonged to the colliery. Proprietors of integrated works however appear to have been reluctant to charge their saltworks an economic rate for coal which they had long been able to deliver to the pans at a modest cost. The substantial difference that could result from underestimating the real cost of small coal can best be seen at Stevenston colliery and saltworks, where this issue formed part of a lengthy and bitter legal dispute between the two partners, Robert Reid Cunninghame and Patrick Warner. While Cunninghame claimed that over a 28 year period to 1798 the pans had made a profit of over £7,864, it was argued by his critics that the costs of mining and delivering the small coal to the pans had been significantly higher than Cunninghame had allowed. It was also alleged that he had sold the small coal to the pans at a rate which was below the local market price, with the result that the colliery should have been credited with an additional £9,751. With costs calculated in this way,the pans would have made a cumulative loss of just under £2,000.[43] Judgement on the correctness or otherwise of Cunninghame's accounting practices however depends to some extent on knowing whether the current market price for small coal, upon which his detractors built their calculations, would have held up if Cunninghame had flooded the local market with the quarter of a million loads of small coal he sent to the pans.

There was even a suggestion that the modest profit of £380 made at Fordel in 1791 might have been inflated by unrealistic accounting. The coal going to the pans had been charged at the 'former price' of 4*s.* 6*d.* per waggon, whereas the current selling price was 7*s.*[44] It was the relatively few saltworks which were not integrated with a colliery which were the most vulnerable however. At their Maryburgh and Craigie saltworks in Ayrshire, the Cadells were not only held to ransom by coal proprietors who would supply them only at the highest prices, but who also refused to sign contracts to supply them with fuel regularly, presumably with the intention of remaining free agents, able to sell to the highest bidder in a fluctuating but generally rising market.[45] On the north east coast, it was only after 1793, when changes were made in the rates of duty payable on coastwise movements of coal, which resulted in an increase in coal shipments at reduced prices from Newcastle, that saltworks could be established. Aided too by a sharp rise in the price of salt which followed from the outbreak of war with France in 1793, the new saltworks which sprang up at Dundee, Usan, Montrose, Nigg, Peterhead and Portsoy, were able to remain in business for some thirty years, in spite of the considerable distance which lay between them and the nearest coalfield.

SALTWORK PROPRIETORSHIP AND ITS OTHER ATTRACTIONS

Over the centuries a distinct pattern of saltwork ownership and management can be discerned. This stretches back to the domination of the industry by church and state in the twelfth and thirteenth centuries, through the phase of predominantly landed proprietorship from the sixteenth century until the eighteenth, and forward from the later 1700s into the nineteenth century when salt pans, if still often part of the stock of a landed estate, were invariably leased to men of significantly lower social status. For the greater part of the period with which this book is concerned then, ownership of the industry was largely in hands of the landed classes. From this source too, emanated much of the industry's entrepreneurial effort, although this could often be indirect, with day-to-day decisions concerning pan management and salt sales being taken without consultation by estate chamberlains and factors. These men were also frequently to be found offering strategic advice to their masters, although at this level of operations the intervention of the latter was more likely. The significance of landed enterprise in Scottish economic and social life from the 1600s, until it began to wane in the early nineteenth century, has, of course, been established elsewhere, notably by Professor T C Smout.[46]

At present it is difficult to say precisely when the religious houses began to give up their direct interests in salt panning. What is probable is that by the fifteenth century, the industry was largely in lay hands. Of a number of pan proprietors listed in the vicinities of Musselburgh and Prestonpans in the 1480s and 1490s for example, only one appears to have had any close religious connections: the occupier of a single saltpan at 'Saltprestone' was Sir Nicholas Flemyn, described as the 'chaplain of the alter of the aisle of St. Mary of Setone'.[47] Other owners or occupiers of pans were variously described as burgesses and, further down the social scale, indwellers. Landed involvement too was in evidence, although in most cases there is no immediate clue as to social status. With its pattern of dispersed proprietorship of small single and double pan undertakings however, the Prestonpans area may not have been typical, although in 1574 there were at least seven different saltpan owners at Kirkcaldy, suggesting perhaps that large-scale landed proprietorship and consolidation of the industry intensified after this date. In the early eighteenth century, when customs and excise records make it possible to identify the names of the proprietors of virtually every Scottish saltwork, Prestonpans stands out from most other salt producing localities, including Kirkcaldy, where the number of proprietors had fallen to two. Many of the fifteen saltworks in the Prestonpans customs precinct were owned or operated by persons of comparatively modest rank. The feuars of saltpans on Prestongrange estate, for example, included John Greig, an indweller, and Richard Henderson, a sailor, who had two pans. Less remarkable was the presence amongst the ranks of Prestongrange salt proprietors of four merchants of indeterminate rank, three of whom were local, with James Donaldson evidently being based in Edinburgh.[48]

In 1716–7, average annual output from the eleven saltworks in the Prestonpans district which were evidently in non-landed hands was not much over 2,000 bushels, compared to a Scottish average of some 4,500 bushels, a figure which is reduced by the inclusion of the small Galloway pans.[49] The few larger works in the Prestonpans precinct, and most important salt panning operations elsewhere in Scotland, were owned and frequently operated by members of the landed classes. Amongst those landowners who had interests in salt were some of the most powerful in the land. The Dukes of Hamilton for example were long-established saltwork proprietors at Bo'ness. The Earls and Dukes of Argyll had pans on the Forth as well as on the Kintyre peninsula. The Maitlands of Thirlestane (the

Earls and Dukes of Lauderdale) had pans on the Forth, near Edinburgh. In Ayrshire, the Earls of Eglinton maintained pans at Ardrossan, while the Hamiltons of Bargany had at least one pan at Shalloch. At Turnberry, the Kennedys of Culzean had a pan and coal mine to supply it with fuel from the early eighteenth century, although compared to most landed enterprises this was a rather modest affair, which produced an annual rental of only 200 silver merks in 1710. Other notable representatives of the landed classes with more sizeable saltworks were the Earls of Kincardine, in the Culross vicinity, and respectively, the Lords St Clair, who had pans at Dysart, and Lord Colville, who operated Craigflower pans, near Culross, in the early 1700s. In terms of the scale of operations, from the second half of the seventeenth century, few could match the Earls of Wemyss, whose salt panning empire at Wemyss and Methil was largely established between the 1620s and the 1650s, or the Earls of Winton, with their extensive range of pans at Cockenzie. At the other extreme was the small but long-lived saltmaking operation at Galdenoch on the Galloway estate of the Agnews of Lochnaw. The family's involvement at this location also went back into the mid-1600s. The list could be extended, but would serve little useful purpose.

The more important task is to account for the intensity of landed interest in salt manufacturing. After all, prior to the first real signs of industrial development in Scotland, with widespread Board of Trustees-inspired investment in lint mills and bleachfields from the 1730s, for example, it was uncommon for large capital sums to be invested in manufacturing units.[50] The fixed capital requirements of domestic industry, in which many landowners were involved, were small. While some large collieries did exist prior to 1707, even in the first half of the eighteenth century, most mines were shallow and small and abandoned with 'comparative frequency.'[51] Saltworks called for what were then unusually large fixed capital commitments. In 1628, for instance, £2,500 Scots was laid out on a pan and panhouse at Wemyss; two years later an additional pan cost 3,000 merks 'in all things, in stone work, timber, iron and lime and workmen.'[52] Nearer the end of the century, as was indicated in Chapter One, it was reckoned that the cost of a single pan and panhouse was in the region of £150.[53] For the larger panmasters, such as the Earls of Wemyss and Winton, who each had more than 10 pans, or the Dukes of Hamilton, who had eight in the 1680s, the total outlay would have been considerable. In 1714, even the average Scottish saltwork, of which there were around 54, comprised three pans. It should be noted too, that neither the figure of £150, nor those from Wemyss, included the costs of girnels and any other necessary ancillary buildings, such as a smithy, stables or housing for the salters, and so it is difficult to estimate the total fixed capital which was required in order to commence salt manufacturing. The cost of a girnel could be in the region of £1,000 Scots at the turn of the eighteenth century. At most works the salt from each pan was stored in a separate girnel. As was indicated earlier, the demands of pan maintenance made for a considerable additional burden on a proprietor's funds.

With the advance of Scottish industrialisation however, the capital requirements of an individual saltwork, and the total fixed capital stock of the salt industry as a whole, paled into relative insignificance. Fearing the consequences of the removal of the duties which had for so long protected Scottish salt, one salt agent in 1788 suggested that it would be reasonable for the state to compensate those saltmasters who were going to be adversely affected at the rate of £200 per pan.[54] It was estimated that there were then 118 'going' pans in Scotland, excluding those in Annandale.[55] On the basis of these figures, the complete panhouse stock in Scotland can hardly have been worth more than £25,000, which was equivalent to the insured value of just one of the country's larger cotton mills in the 1790s.[56] On the other hand, the sum of £25,000 does not appear to have included the industry's ancillary buildings, and furthermore, it was based on perceptions of its condition in the rather dismal early 1780s. Nonetheless, as it was reckoned that there had been around 164 pans in Scotland at the beginning of the century, it seems likely that the capital value of the industry had declined in absolute terms. Later though, during the boom of the early nineteenth century, over £2,000 was spent on building just two pans at Brora, and, as was indicated earlier, several new works had been established after 1793.

Even so, the industry was unquestionably experiencing long term relative decline, as can be seen in the valuations of individual saltworks. In 1813, for instance, the ironwork (which was by far the most expensive element in the stock of a saltwork) belonging to Sir John Callendar's three going pans at Prestonpans was valued at just over £362. An adjoining horse mill was worth a mere £29.[57] In 1844 the same works was judged to be worth only £720. At Prestonpans and elsewhere, by mid-century, saltworks were being let for £50 or £60 per annum, compared to the hundreds of pounds the owners of the largest works had been able to command up to fifty years earlier.[58] Gone forever were the heady days of the mid-1790s, when competitive bidding had raised the annual rental for the St Clair's saltworks from £100 to £200, with the owners having the option of taking four-fifths of the profits instead of the fixed rental.[59]

What, however, accounts for the intense interest which owners of coastal estates showed in salt manufacturing prior to the middle and later decades of the eighteenth century? A principal reason, of course, was the interlocking relationship of salt and coal, previously discussed. Salt sales, however, sometimes did more than simply make coal mining operations profitable, although this in itself was a major contribution to estate development. In the boom years of the seventeenth century, and even beyond, income from salt could form an indispensable part of an estate's income. At Wemyss, for example, in 1634, David, the second Earl, hoped that with 'God's help', the pans would make a profit of 4,000 merks, of which 1,000 were to be used to purchase fresh meat for his household.[60] It is significant that Earl David paid none of his staff any more than the 100 merks that went to Alexander Wemyss, 'for keiping the salt of my pans'.[61] On the Prestongrange estate of William Morison in 1716–7, income from

9 Pair of surviving panhouses, Prestwick, 1986. These buildings, which are currently used as greenkeepers' storehouses on St. Nicholas golf course, were probably constructed in the 1760s, and are referred to in the text as Craigie or Maryburgh saltworks. They are certainly the most complete upstanding remains of the industry on the west coast. The salters were housed in the upper storeys. Note the sea water inlet pipe on the seaward gable of the nearest panhouse. Photograph by C. A. Whatley.

salt sales accounted for around 63 per cent of non-agricultural income, and was thus much more significant than sales of coal, glass bottles or cloth.[62] Yet cases such as this are very much in a minority. It is in less obvious ways that the impact of salt sales upon the landed estate was felt.

As is well known, by far the greatest part of estate income in Scotland came from rentals of agricultural land and the sale of crops and livestock. A large proportion of this, however, was often received in kind.[63] Salt sales though were largely transacted in cash. In spite of the emergence of more sophisticated commercial arrangements, through the use of bills of exchange for instance, much routine daily business, conducted at a level just out of sight of a posterity which relies so heavily upon surviving written sources for its understanding of the past, could only be transacted in money. Yet in Scotland, it is widely accepted that there was a chronic shortage of specie.[64] Thus, given the universal requirement for salt, for landowners fortunate enough to be the possessors of coal or peat bearing estates bounded by the sea, salt manufacturing could go a long way towards plugging their liquidity gap. Interest in the extractive and manufacturing aspects of estate improvement, of which salt making was a part, received a general stimulus in the later seventeenth century from the low prices obtainable for grain,[65] and a more positive incentive from the buoyancy of the salt market.

During the crisis years for the Scottish economy, of the 1690s and early 1700s, there is evidence that the importance of salt sales may have been raised to altogether new heights. This is best seen on the Duke of Hamilton's extensive estate of Kinneil. Problems which had been emerging for some time, had become acute by 1704, when rents were long overdue, crop yields were disappointing and much estate work remained undone owing to the consequent shortfall in income. Virtually the only sign of expansion lay in the direction of salt manufacturing, which the Duke's agents were attempting to extend. Even though volume sales would probably only be achieved if the price of a boll of salt was reduced by 2*s.* Scots to 36*s.*, for the chamberlain, Daniel Hamilton, such a concession was essential, 'for money must be had som way', he wrote to the Duke, 'and I know of no other.'[66] Daily running costs on the estate could not be met unless the shortfall in agricultural rents was compensated for by salt sales.[67] As will be seen in Chapter Four, the political consequences of these circumstances were of the greatest long term significance for the industry.

Even in less obviously difficult times, cash generated from salt sales clearly played an important role in the affairs of saltwork-owning estates. In the early eighteenth century, for instance, the St Clairs evidently considered saltwork income to be an integral part of their household finances. Sums were drawn directly from saltwork funds to pay for a variety of expenses which were vital to sustain a landed family in reasonable comfort. Paid for in 1713, for instance, were wine, ale, vinegar and salmon and ling for the house of St Clair. The kitchen 'poats and pans' were tinned for £14 2*s.* Scots, a 'sadler' was paid £3 Scots for mending the saddles and a further £3 Scots was laid out for 'Mending the Cloak in the Dyning Room'. The wages of household staff too were paid from the same source, as were the costs of their 'britches' and other clothes. £33 Scots, for example, was paid to Dugall Mackom, the cook, for half a year's work, whereas the yearly fee of a 'servant' was £48 Scots. Personal expenses too were provided from the monies raised from salt sales: amongst the items recorded were travelling expenses for Lord St Clair and his staff, payments to Robert Stewart, for dressing wigs, and to others for writing paper and books.[68] Most telling of all perhaps was the £2 10*s*. used by the factor 'when at Edr borrowing money from the Bank for payment of My Lords Bills'. In this way, saltworks could perform a similar function to that which agricultural income performed on inland estates such as Glamis, in Angus, where money derived from agricultural activity was used to purchase small scale household goods.

Saltwork cash too facilitated the sinking of new coal pits and the driving of new levels in existing mines. During such periods, which could last for several months or even longer, as coal was generally not being put out and sold, colliery income was often much-reduced or even non-existent, and difficulties could be experienced in obtaining the cash required to pay sinkers and retain underemployed coal hewers. At a time when colliery labour could only with difficulty be recruited into the coal industry, it was unwise for coal proprietors to depend solely on the law and customs which bound colliers and their families to their estates to ensure their labour supply. Unpaid colliers tended to become 'turbulent', or 'mutinous', and drift away in search of alternative employment.[69] It was not always easy, or cheap, for coalmasters to get their absconding colliers back. In these circumstances, the possession of an adjoining saltwork was a major boon, as in 1722 when attempts were being made on the Wemyss of Bogie estate to find 'great' coal. The expense of sinking, it was reported, would be diminished, 'with what money Mr Wemyss will get in for Salt'.[70]

Finally, given the importance of salt in pre-modern Scotland, both to preserve and to season foodstuffs, it seems possible that the maintenance of an estate saltwork may have been judged to have been a highly practical way of avoiding the salt shortages which could occasionally cause great distress for those less fortunately placed. Certainly most landed saltwork proprietors relied on their own salt to some extent for their household requirements. Salters at Elphinstone pans for instance could expect that some of their output would be taken 'to the pleace for the families use' during the early eighteenth century.[71] During the 1690s, salt, 'for ye use of the Duke of Hamiltons family', was taken regularly from the pans at Kinneil, to serve the household on that estate, as well as to Hamilton Palace itself, and the family's apartments in Edinburgh. The practice continued into the eighteenth century, although usually on a more formal basis. From 1710, the lessees of Sir Archibald Kennedy of Culzean's pan undertook to provide Kennedy with four bolls of salt each year for his 'house', in addition to their cash rental.[72] As late as 1762, Sir Andrew Agnew of Lochnaw required a new tenant of the farm and saltpan at Meikle Galdenoch to 'provide the family of Lochnaw with what Salt shall be made use of for the family', although in this case, the tenant, Patrick McNeilly, was to be allowed a deduction from his cash rental at the rate of £8 Scots for every boll of salt with which he supplied Agnew.[73]

NON-LANDED SALTWORK PROPRIETORSHIP AND LESSEES

While in the seventeenth and first half of the eighteenth centuries the industry's larger works were mainly built and operated by substantial members of the landed class, as has been seen in the case of Prestongrange, saltwork proprietors and lessees could also come from other, lower, social groups. In Chapter One, attention was drawn to the prevalence of tenant farmers as salt makers in Galloway. They were also to be found operating pans elsewhere in the country, as at Ardrossan in 1778, when William Bone, of Mid Moncur farm, leased the two pans belonging to the Earls of Eglinton.[74] Unfortunately though, little evidence from other parts of the country, which would make possible a systematic analysis of the social background of non-landed saltwork owners and lessees, has survived. An additional problem is to determine whether a lessee or partnership was interested primarily in salt, or whether they should more accurately be described as coalmasters. There were comparatively few cases where this is absolutely clear, as it was at Fordel in 1746, when Hugh Wilkie took a three year tack of three saltpans, but had nothing to do with mining coal, which was to be supplied by Sir Robert Henderson.[75] Given the way in which the two activities were so closely linked, the roles of coal and saltmaster were clearly often inseparable, and further information on their background can therefore often be drawn from the work of historians of the coal industry.[76]

In those cases where it seems reasonable to assume that salt was what lessees were principally interested in, what the available data indicates is that there were many routes into the salt industry. People who described themselves as 'merchants' however, appear to have been the most prominent lessees of saltpans. Too rarely, however, is it possible to say what status such persons held within the merchant communities from which they were drawn. Sometimes, they were evidently men of some substance, with fairly ready access to capital. From 1709, for instance, Hugh Hamilton, an Ayr merchant, along with his partner Patrick Young, a writer in Maybole, undertook to bear the cost of building a panhouse, bucket pot, girnel, coalyard and salt officers' houses on land belonging to Sir Archibald Kennedy of Culzean. At the end of their 19 year lease they had the right to remove all the ironwork, but 'no wayes' could they take any stone.[77] In 1728 the pan and its adjoining colliery were let to two less well-endowed merchants, John Teugh and John Ross, of Girvan.[78] Further north along the Ayrshire coast, at Ardrossan, the Earl of Eglinton's pans were at one time let to William Orr, a merchant from Kilbirnie, who, around 1759, with William Mitchell, a Paisley merchant, took a lease of the Duke of Argyll's saltworks and colliery at Marypans, near Campbeltown. Unfortunately, little more can be said about their backgrounds. They were, however, joined by Samuel Mitchell, a former shoemaker, who had for a short time been involved with Orr at the saltworks at Ardrossan.[79] In 1754, the pans at

Inverkeithing were let to John Henderson, who was described simply, and unhelpfully, as a merchant. By 1761 the same pans had been taken over by Ebeneezer Deas, 'Late Keeper of the Toll-bar at Inverkeithing'.[80]

During the second half of the eighteenth century, direct landed involvement in salt began to recede, as it did to a lesser extent in coal. The movement was not always outwards. Although for decades the Hendersons of Fordel had leased their pans at Inverkeithing to a variety of tacksmen, in the 1770s and 1780s Sir John Henderson not only took a direct interest in their management, but also played a leading part in the activities of the Salt Association. In general, however, the market for salt was flat, and it was rentals from coal mines, which were being freed from their former dependence upon salt, that promised the greatest returns. The final burst of landowner interest in saltpan construction and operation came during the second half of the 1760s and first years of the following decade, although landed enthusiasm was shared by others. Demand for salt strengthened markedly during the inflationary boom which was to be cut short in 1772 with the collapse of Douglas, Heron and Company, the Ayr Bank.[81] Signs of an upturn in the salt trade had been seen earlier. In 1763, for example, William and Alexander Mure of Caldwell had been granted permission to build a new saltwork and ancillary buildings in the lands of the burgh of Prestwick.[82] The sharpest burst of activity came at the end of the 1760s though. A 'company' from Portsoy restarted salt manufacturing at Brora. Five ruined pans at Grange were recommissioned by the Cadells, and John Roebuck built his 'monster' pan at Corbyhall. Planned earlier, but commencing production in 1772, Sir John Anstruther's nine-pan complex at St Monance was to become the third largest saltwork in Fife.[83] He was not alone in this venture however, which for a time was in the hands of the Newark Coal Company, of which few documentary traces appear to have survived.

Although, as will be seen in the following chapter, the landed retreat was not immediate, salt manufacturing gradually fell into the hands of men of lesser rank. The by now relatively small capital and operating requirements of the industry opened it up to a host of minor merchants such as James Dickson of Montrose, who converted to saltpan use the buildings of the Greenland Whale Fishing Company. The lessees of Dysart saltworks from 1797 were William Fleming, George Rutherford and James Normand, shipowners.[84] The suggestion here that salt manufacturing was being perceived as a logical means of diversifying and extending existing business interests, rather than a sufficiently rewarding activity on its own, is further supported by the short-lived Dundee Glasswork Company's apparently unsuccesful attempt to make salt from the unsuitable water of the river Tay.[85] Also coming in were men from a wide range of backgrounds even lower down the social scale, such as Hugh Currie, a smith, who took over and rebuilt the Earl of Eglinton's pans at Ardrossan.[86] It was presumably people of this sort who led the short-lived saltpan building boom at the beginning of the nineteenth century. However, as in the main, the larger saltworks at last became merely the appendages of collieries, so the account of their ownership and management merges with, and indeed is superseded by, that of coal mining.[87]

4

THE POLITICAL ECONOMY OF COMMON SALT

INTRODUCTION AND DEVELOPMENTS TO *c*.1661: HORIZONS ABROAD

> People of the same trade seldom meet together, even for merriment or diversion, but the conversation ends in a conspiracy against the public, or in some contrivance to raise prices.[1]

Adam Smith's observation of the tendency of businessmen to act in concert, preferably unseen, for their own benefit, is probably no more apposite than when it is applied to the leaders of the Scottish salt industry. Indeed, it seems reasonable to speculate, although it cannot at present be substantiated, that Smith's views were at least in part based upon what he may have learned of the secretive activities of the Scottish saltmasters during the periods he spent in Kirkcaldy and the east of Scotland.[2] Examples of employers' combinations in pre-Victorian Scotland have been discovered and discussed. The Glasgow Coal Company, for instance, emerged around 1790, with the aim of controlling members' coal output and fixing prices. Under several guises, it continued to operate until 1817.[3] The Scottish ironmasters too, acted in unison in the early nineteenth century.[4] South of the border, much attention has been paid to several powerful monopolising schemes which grew out of the coal industry of the north east, such as the seventeenth century Company of Hostmen, the 'Grand Alliance', which emerged in 1726 and lasted into the nineteenth century, and the 'Limitation of the Vend', which appeared in 1771.[5]

It is with the the latter associations that the Scottish saltmasters are best compared, both in terms of the longevity of their interest in collusion and the success with which they conducted their operations. Although there is no evidence of continuous formal association, there are strong indications that an identifiable salt manufacturing 'interest' existed and that informal contacts were increasingly maintained between the leading saltmasters from the late sixteenth century. Out of these developed common business strategies which continued to be adopted by the bulk of the Scottish saltmasters until the later eighteenth century. On at least three occasions, in 1636, from 1670 until 1673 and from 1773 until the 1790s, separate types of formal associations of saltmasters were to be found, with the latter two unambiguously intent on feathering their adherents' nests at the expense of Scottish consumers.

Ebbs and flows in the political fortunes of the Scottish landed classes in the sixteenth and seventeenth centuries notwithstanding,[6] the saltmasters, individually and in concert with others, clearly had little respect for the central authorities and frequently flouted the wishes of both Church and State. Privy Council decrees and Acts of the Scottish Parliament were ignored, with the central authorities simply lacking the means of implementing their policies regarding salt. On other occasions, when state decisions which may have been harmful to their interest looked like being implemented, they tried to ensure that such policies were sufficiently modified in their favour. In this they were assisted by the close links most of the leading saltmasters developed with central government, sometimes as part of it, after the Restoration. When in the eighteenth century the fiscal arrangements concerning salt favoured the established saltmasters, they did all they could, largely behind closed doors, to resist the changes which were proposed by more progressive economic elements. Scottish landowners generally may have played a critical positive role in the modernisation of the Scottish economy: such a case,[7] however, cannot without considerable qualification, be made where their interests as saltmasters were concerned.

As early as the 1520s and 1530s, there is clear evidence that the salt manufacturers of Scotland preferred to follow their instincts as private capitalist producers rather than serve the public good. In May 1526, the Lords of Council were aware of complaints that there was an 'exhorbitant dearth' of salt in Scotland and that the 'lieges can nocht be servit apoune ane competent price as has bene usit in tymes bigane'. A ban was therefore placed upon the export of salt and in 1527, to further conserve supplies, none was to be sold to 'strangers'.[8] The Lords' decrees, however, evidently fell on deaf ears, as in 1531 it was reported that their proclamations had had little effect.[9]

State regulations governing the sale and export of Scottish salt were similarly ignored in 1573 and 1574, as opportunities for overseas sales opened up with the collapse of the Netherlands' salt refining industry and a drop in the production of French salt.[10] Certain panmasters on the Forth, it was alleged in March 1573, were selling salt in Scotland 'upoun grit and exorbitant prices, in manifest hurt and prejudice to the common weill'. What emerged from the events of these years, which exposed the state's powerlessness to bring to heel the saltmasters of the Forth, was an acceptance by the authorities that compromise would be necessary if there was to be any likelihood of the national interest (a certain supply of Scottish salt for the 'lieges' at affordable prices) being secured. With the appearance of a delegation from Prestonpans before the Privy Council in December 1574, came the first certain sign that the saltmasters had begun to organise amongst themselves. Pleading that if they were forced to depend upon the 'small pryces' to which a recent Act of the Scottish Parliament had restricted them in the limited home market, they would not be able to cover their operating costs. They insisted therefore that they should be allowed to continue to export salt.[11] In a series of virtually identical deals struck with groups of saltmasters from each of the principal producing districts, licences to export salt were granted on condition that from each 'gangand' pan, crown representatives were to be supplied with three bolls of salt weekly. This they were to sell to domestic consumers at 10*s*. 4*d*. Scots per boll,

10 Portrait of David, 2nd Earl of Wemyss (1610–1679), attributed to Schunemann. Wemyss established one of the country's largest salt manufacturing empires in the mid-seventeenth century, at Wemyss and Methil in Fife. The Wemyss family played a leading part in the industry thereafter, politically as well as economically. In a private collection.

3*d*.Scots more than the pan proprietors were to receive.[12]

Uneasily then, the conflicting needs of private capital accumulation and of the paternalistic state to ensure justice and fair play for its inhabitants, were met, at least on paper. Further conflicts did arise, however, and a series of monopolies granted by the Crown to individuals and partnerships who undertook to improve the manufacture of salt in Scotland cannot have been viewed favourably by the established saltmasters. Indeed, there were occasions when their sensitivies were clearly recognised. In 1588, for example, James VI awarded extensive rights to Eustachius Roghe in return for his radical proposals for making more and better salt, cheaply, in works he was to erect near Leith. At the same time, however, it was open to 'ony personis' to 'follow furth their auld accustomit fasshion of making salt in the auld maner of fornaceis and be na uthir new forme nor devyse'.[13]

The imposition of the export duty in 1584, referred to in Chapter Two, cannot have been welcomed either. Although it appears to have been collected with some success until the close of the century, in 1618 the Privy Council heard complaints about the 'verie grite' abuses of the fiscal regulations by 'maisteris, awneris, and fermoraris' of saltpans as well as coal mines above Queensferry. These had allegedly been taking place for some time.[14] Even the threat of invasion in 1627, which drove the Privy Council to take urgent steps to raise funds to pay for the improvement of the fortifications of the Forth, appears to have failed to persuade the saltmasters that they should subordinate their own interest to that of the defence of the realm. Both the owners of coal mines and saltpans, their industries being judged 'the most lucrative...of that part of the coast' and therefore best able to pay for its defence, paid little or no attention to the Privy Council's imposition of 2*s*. Scots payable on each exported ton of their respective commodities.[15]

Nothing, then, was to be allowed to inhibit the saltmasters from maximising the output potential of their pans or from reaping the full benefits of the higher prices which overseas sales brought during the prosperous decades of the early seventeenth century. In 1606, the Act which debarred the salters (and colliers) from moving freely to new employers was passed. Further evidence that the saltmasters perceived themselves as having a common interest comes from 1636, when, in order to secure certain privileges, they obtained from the Privy Council the right to establish an 'incorporation of Salt-maisters'.[16] Unfortunately, nothing further is known about this body, for which the saltmasters may have had little necessity during the boom conditions of the later 1630s. Attempts from 1640 by the Church and its zealous henchmen in the saltmaking localities to stamp out the fires of those salters who continued to make salt on Sundays evidently had little impact. Favourable market forces were to be allowed to rule, rather than the laws of God's earthly representatives. During less propitious times, the strategy of open defiance was replaced by special pleading. Thus in 1645, during a brief period of dislocation for the exporters of both coal and salt, John, Earl of Wemyss and Lord Elphinstone, on behalf of most of the country's coal and saltmasters, took the opportunity of asking the

11 Portrait of Alexander, 2nd Earl of Kincardine (d. 1681), attributed to Jan Mytens or Hanneman. Has the dubious honour of bringing the question of salt supply in Scotland to the forefront of Scottish politics during the reign of Charles II. From 1671 until 1673 he had the sole responsibility for (and the profits from) the distribution and sale of virtually all salt made and sold within Scotland. In the collection of the Earl of Elgin.

Scottish Parliament to relieve them of the burden of a tax which had been imposed on them in 1641 in order to cover part of the cost of maintaining the Scots army.[17]

SECURING THE DOMESTIC MONOPOLY

With the disappearance of former trading opportunities overseas from the middle decades of the seventeenth century, the former inclination of the saltmasters to act together became a pressing necessity. Uncertain unions of the previous seven or eight decades were replaced by a hitherto unwarranted consistency of purpose: at stake was if not the survival of salt manufacturing in Scotland, at least its continuation on the scale those enagaged in it had become accustomed to. As was seen in Chapter Three, the maintenance of coal mining in the coastal districts was also thought to depend upon a thriving salt industry. Generally, although not altogether dismissive of the home market since the early 1570s but evidently prepared to leave others to provide for Scottish needs, the saltmasters now strove jealously to secure the domestic trade for themselves. While for decades it had generally suited the saltmasters to disregard the state—the failure of several of their number, including Alexander Hamilton of Kinglass and Alexander Bruce and John Adie of Carriden, to collect the levies on coal and salt in 1627 had caused them to be summoned 'on pain of rebellion' to appear at Holyrood and explain themselves—they were soon to become the firmest advocates of state policy towards salt. First, however, they had to ensure that policies which served their interest were adopted. The restoration of the aristocracy to power in Scotland in 1660 made their task relatively easy.

The first indication that the saltmasters had embarked upon a protectionist tack came in 1661, when in an Act designed to raise £40,000 Scots for the Crown, Parliament favoured Scottish producers by levying significantly lower duties on native salt than on imported varieties. Each Linlithgow boll of Bay and of refined salt, for instance, was to be taxed at 40*s*. Scots; only 15*s*. Scots was to be imposed on a boll of 'inland' salt.[18] A decision by Charles II in 1665 was even more specific about and helpful to the Scottish industry. Sir Walter Seton had led a protest claiming that 'great quantities of forrain salt' were being 'consumed for the use of domestick families'. This, it was argued, was detrimental to the revenue and a 'discouragement' to the saltpan owners which would lead to the 'ruine' of the salt industry and the 'poor people ther'.[19] Dressed up in terms of the interest of Scotland generally and the need to stem the outward flow of specie used to pay for foreign salt, a decision was then taken by the Privy Council to impose a further tax of £12 Scots per boll of any salt, 'not of the product and manufactory of our dominions of Scotland, England and Ireland', except where this was to be used for curing flesh for export.[20] The specific aim of the measure, of course, was to support the allegedly 'ruined' collieries and saltworks of the Forth, the proprietors of which, as landowners, were almost certainly feeling the pinch of low rents during the early stages of a prolonged period of reduced corn prices.[21] Thus was established a policy of protection for the Scottish salt industry, which, further strengthened, was to remain in force, carefully guarded by several generations of saltmasters, until the early nineteenth century.

The sharp shift in official attitudes to salt and the degree of protection which the saltmasters achieved in Restoration Scotland brought an outburst of protest from those who were forced to use their unpopular product. Nevertheless, as Aberdeen Town Council and a succession of disgruntled salt users were to discover, the interests of consumers, both domestic and industrial, were henceforth deemed to be secondary to those of the salt producers. Justifiable claims that the measure of 1665 had been designed to assist 'certane noblemen and others in the south for ther awine particular ends' and that Scottish salt transported by sea for any distance could not be used as it was 'in hazard of being dissolved into water if season prove either windy or rainy',[22] made little impact in government circles in Edinburgh. For almost a century and a half, consumer complaints about the price, quality and supply of the Scottish-made salt they were forced to use, were, as often as not, met with official responses which differed little from the statement made by the Privy Council in 1681: 'the salt manufacture of this kingdom will abundantly serve for all domestick uses'. Equally ineffectual were objections from the fishing interest about the 'vexations' caused by the requirement of merchants to enter imported salt officially, or from merchants that 'whole tracts of the country' relied on imported salt and that in its absence they would be reduced to 'great extremity'.

There were, however, political limits to how far the saltmasters could indulge their new found legitimacy. From 1670 until 1673, the activities of some of their number were subjected to increasing scrutiny and in November of the latter year, they found themselves near the eye of a storm of political unrest which blew up around the Duke of Lauderdale, Charles II's Scottish Commissioner.

Great hostility towards the measures of 1665 was felt by merchants, fish curers and the representatives and inhabitants of particularly the royal burghs which lay some distance from saltpans. Not surprisingly, they were unconvinced by the arguments advanced by the country's leading saltmasters, about the sufficiency of Scottish salt, even for fish preservation,[23] or by their suggestion that the interest of their opponents was 'private and inconsiderable' and more than offset by the 'publick good of the Kingdom', including the 'Subsistence of many poor Families & persons' and the support of many 'necessarie' collieries, which had resulted from the decision to discriminate in favour of Scottish salt.[24] Yet in spite of the unpopularity of the direction in which official policy towards salt was going, in September 1669, several saltmasters, including the Earls of Kincardine and Rothes, had made proposals to Charles II, which would place even stricter controls on the importation, distribution and sale of salt in Scotland.

By April 1670 their scheme had been approved by the Earl of Lauderdale and orders were issued by the Lords of the Treasury, for the management of what was described as the 'preemption' on salt. With overall policy in the hands of leading Scottish politicians who had major interests in salt manufacturing, principally the Earls of Tweeddale, Kincardine and Rothes, at ground level the scheme was operated by Sir Patrick Murray, Robert Milne, provost of Linlithgow, and Alexander Milne, a merchant there.[25] Whatever the benefits which might accrue to the Crown in the form of revenues, it was the saltmasters who were to be the main beneficiaries. Henceforth, all salt, whether made in Scotland or imported from France, Portugal or Spain, was to be bought up and then sold at prices determined by the Treasury Commissioners. Critical to the scheme's success, from the saltmasters' point of view, were two factors. First, they now had a secure market for their salt—the agreement was to last seven years, or as long as the king saw fit—and secondly, the 'Salt Managers' were under strict instructions not to allow any foreign salt to be used in Scotland except for the preservation of fish.[26] Apart from the managers, no Scottish subjects were to be allowed to import salt, while 'strangers' had to obtain a warrant before selling any salt to Scots.

The 'publick servants' who were charged with implementing the scheme, however, seem either not to have carried out their responsibilities or to have been too anxious for 'private gain'.[27] As a result, in 1671, Lauderdale and his aides, credited with freeing Scotland from the 'bonds of old monopolies',[28] created a new one, by placing the sole responsibility of managing the salt trade in Scotland and collecting the customs and excise duties due upon it, in the hands of the Earl of Kincardine, in return for an annual rental of £2,000 Scots. It was a mistaken decision. Simmering discontent about salt supplies within Scotland boiled over into outright opposition to Kincardine and the saltmasters. Whereas beforehand the scheme, nominally at least, had remained under royal control, it had now clearly fallen into the hands of a 'private person'. Complaints, according to Sir George Mackenzie of Rosehaugh, 'throng'd in daily', while others were aware of 'constant clamours', about the price of salt, which had allegedy quadrupled, and of the inability of the inhabitants of Galloway, the west coast and the Highlands to obtain salt at any price. Some of them, Mackenzie claimed, were 'forced to use salt water, by which many of them died as of plague'.[29]

Rising hostility might have been tolerable had the scheme's twin aims been met. Yet royal revenues do not appear to have been enhanced by anywhere near the £18,500 Scots hoped for by the promotors of the 1670 preemption, although Kincardine seems to have done well from it.[30] The high price of Scottish salt had reduced its sale in its last significant market outwith Scotland, England, and even at home, its consumption had fallen, with a far greater quantity of salt 'lying in the salt masters hands than found at any tyme befor the preemptions'.[31] By August 1673 the king was being advised by the Exchequer to drop the scheme, which had become, 'grievous to the people and so unprofitable to your Majestie'. In order to encourage the saltmasters to agree to abandon the agreement they had come to earlier with the Treasury Commissioners, of which there were still five years to run, 'all care' was to be taken to see 'that their salt is only imployed for domestick uses and they can export also profitably'.[32] In order to achieve this, all Scottish salt was to be free of excise duty: all foreign imports were to be excised at the rate of 40*s*. Scots the Linlithgow boll.[33]

Opposition to Kincardine, though, was not based wholly on concern about the supply and price of salt and was bound up with wider issues and the shifting sands of Scottish political affiliation in the later seventeenth century.[34] Kincardine's position, however, was dependent upon Lauderdale, against whom considerable ill-feeling had developed by November 1673, when the Scottish parliament reassembled.[35] The issue of the preemption on salt was at the top of a list of specific complaints of Lauderdale's opponents, which also included the import duties which had been imposed on tobacco and the ill-effects of the gift of seizures of smuggled brandy which had been given to Lord Elphinstone. Not surprisingly, the preemption had few supporters. The Earl of Tweeddale, who had certainly been in favour of the royal preemption, but who with the Duke of Hamilton was leading the anti-Lauderdale forces, became 'the great haranger against it'.[36] Rothes too, now in Hamilton's camp, turned his back on his former ally. Within days, Kincardine's monopoly was broken and the scheme laid to rest.[37]

Although their hopes of obtaining greater rewards than salt could ever bring had apparently caused something of a rift amongst the ranks of the Scottish saltmasters, they quickly regrouped and continued to agitate in defence of their industry, aided of course by their proximity to the levers of state power. Although the home market was being secured, there was still concern about the size of the import of foreign salt and the uses to which it was being put. In 1681, the saltmasters, now led by the Duke of Hamilton, succesfully urged, amongst other things, that further restrictions be placed on salt imports. French salt, it was proposed, should be allowed to enter Scotland only if it was used for curing salmon, while Spanish salt was to be solely for the preservation of herrings.[38] The saltmasters too benefited from the policy of 'root and branch' protection of native manufactures which was adopted by the state in 1681.[39] Henceforth, Parliament declared, 'all the Salt works of this Kingdom...free and publick Manufactories'.[40] Further support for the industry, along with coal, was given by an act which relieved both from bearing any part of the 'supplie'. Thus, mobilising opposition to the saltmasters was increasingly difficult for burghs such as Aberdeen, who had rejoiced at Kincardine's downfall in 1673, but who likened the 1681 measures to the earlier preemptions and regretted that, 'some cheife burrowes ar as vehement for this monopolie of salt as the nobilmen concerned and it is to be feared ther may be mannie such amongst the small burrowes who certenlie will be influenced be these nobilmen.'[41]

Even though their grip on the Scottish domestic market was tightening, two other problems caused considerable concern for Scottish saltmasters in the decades immediately preceding the Union of 1707. First, as was seen in Chapter Two, there was the hardly discernable, but nevertheless growing, threat from England. Indeed, a major reason for the preemption schemes, according to Sir George Mackenzie, was 'that our salt might not be ruin'd by the salt made in England.'[42] By the 1680s, quantities of Cheshire's recently discovered rock salt were evidently being shipped to Scotland.[43]

Intimately linked with this concern was the fear that 'salt upon salt' would be succesfully made in Scotland. 'Salt upon salt' was high quality salt which had been refined by methods which had long been used by the Dutch.[44] This process could also be carried out in Scotland, using her ample resources of sea water and small coal. The basic raw material, however, was either Bay salt or English rock salt. Schemes for using the former had been proposed, as in 1621 when the Privy Council granted Robert Menteith the right to 'the making of salt upoun salt'.[45] However, the appearance of rock salt, which was cheaper than foreign salt, added to the anxiety of a number of the established saltmasters. 'A number' is the most appropriate term here, for there is a suggestion that this concern may not have been uniform. There was apparently something of a division in the saltmasters' ranks, between a majority who wanted nothing to do with schemes which would involve the free importation of foreign and rock salt and those who were at least prepared to consider improving the quality of their product and attempting to satisfy fishery demands. On the other hand, what appears to have united them were proposals for any monopoly rights, which might ultimately have threatened their own ascendancy.

However, what is clear is that considerable hostility to the introduction of new methods was generated in 1696, when at least two schemes for making salt upon salt came before Parliament. While many saltmasters appear to have been unhappy that William Erskine's proposal was approved, it was towards the African Company's proposal to make 'new fashion salt' that their anger was principally directed. Geared towards the production of fishery salt (it was hoped to send salt fish, along with other commodities, to Africa), the scheme, which was for a monopoly of the manufacture of salt made in the way the Company proposed, included the suggestion that if necessary, any lands 'that lyes near the sea' between Alloa and Culross, where the refining works was to be sited, should be compulsorily purchased.[46] If approved, the Earl of Arran was urged to believe, it would 'destroy all the Coal and Salt masters Trade in Scotland'. It was a stab in the back for many of the backers of the African Company who also had interests in the coal and salt industries: 'I don't understand the African Company', wrote one member of the ducal Hamilton family, 'for I thought they were to trade to the Indies and not sett up to ruin manufactories that are established by law'.[47] However, although strongly supported in Parliament, concessions were obtained: monopoly rights were not granted to the African Company and land could only be be purchased by it where ownership was doubtful or where the property was 'incumbered'.[48] While it is uncertain if the Company ever made any salt, their more grandiose plans vanished along with so much else that had been hoped for from the ill-fated 'Company of Scotland trading to Africa and the Indies'.[49]

If this was the most serious threat to the marine salt industry in the period, the saltmasters still had to dispose of the danger from English salt. Concern was such that in 1701 the coal- and salt-masters made an unsuccessful attempt to obtain parliamentary approval for an uncompromising and wide-ranging measure which, amongst other things, would have prohibited the importation of 'small White Salt from England or elsewhere'.[50] They did far better in 1705 however and a blanket ban on all English salt was imposed.[51] After almost five decades of struggle, the saltmasters, with their allies in coal, had managed to eliminate all legal competition from the domestic market for salt in Scotland.

MAINTAINING THE MONOPOLY, *c*.1707–1773

As far as the economic dimension of the Union of 1707 is concerned, it has become commonplace for historians to think in terms of a 'common market': economic salvation for the Scots, it is argued, lay in the larger markets of England and her colonial empire.[52] The saltmasters too, it has been suggested, were also firm supporters of a measure which would allow free entry of their salt to England.[53]

Yet the argument that the saltmasters' interest lay in the direction of England is based upon the belief that the industry was declining in the late seventeenth century and that its leaders clutched at the 'straw' of union, not because it would actually save them, but because the last prosperous period they could remember had been the time of the Cromwell's rule.[54] Yet, as was demonstrated in Chapter Two, in the two decades which preceded 1707, the salt industry appears to have been reasonably buoyant. Its strength however, was based increasingly on the home market. Further, as has been argued in the previous section of the present chapter, the saltmasters' hold on the home market was underpinned by a series of legislative steps which had been taken to exclude competition, including that from England. Scottish success abroad, when it occurred, was largely due to temporary dislocations in the trade in the preferred varieties of salt.

It seems highly unlikely, then, that many proprietors of Scottish saltworks would have been prepared to throw aside the armoury they had spent half a century acquiring. At a time when several weaknesses in the Scottish economy were being exposed it appears more likely that the saltmasters would have sought to maintain what they had, rather than opt to take an extremely risky leap into markets from which they had been excluded since 1680. An acute awareness of the present, rather than what must for many have been second-hand memories of a situation which had prevailed several decades earlier, guided the business strategies of Scotland's saltmasters in the last months of 1706, when the duties to be imposed on salt in post-Union Britain were being debated. In Chapter Three, attention was drawn to the weight of interest that hung on the continued success of salt.

And although it is largely by inference that the aims of the saltmasters during the negotiations which led to the Union can be discovered, their principal objective does appear to have been to secure Scotland, for themselves. As was pointed out in the Introduction, salt was a major issue in 1706: during the final stages of the debate on the Articles which were to form the basis of the union, in December 1706, the Earl of Mar wrote to Sir David Nairne: 'Today we were on the 8th artickle in relatione to salt...I told you always that this was the artickle I was most affraid of'.[55] Several amendments were made to the Article, which was concerned with salt and fish. Reflecting the recent division in the saltmasters' ranks over the question of salt upon salt, there does appear to have been an unsuccessful attempt by an unknown number of saltmasters, in alliance with the fishing and fish curing interest, to have the freedom to make such salt included in the Union settlement.[56] However, there is no evidence of any objections being made to the clauses which insisted that Scottish salt entering England should be liable to pay the same duties as were payable on English salt. The duties on foreign salt entering Scotland were also the same as those charged south of the border. The issue which aroused the greatest passion was the level at which duty was to be charged on domestic salt. Unfortunately for posterity, Daniel Defoe declined to report most of the speeches which were made on the subject, or the names of those who made them, 'for the sakes of the Gentlemen that shewed themselves a little out of Temper here.'

To demand, as many Scots parliamentarians did, that Scottish salt should be excluded from taxation altogether, on the grounds that Scotland's poverty made her inhabitants more heavily dependent upon salt than the English, was contrary to the principle of free and equal trade which had been agreed in the IVth and VIth Articles. And yet after a heated and emotional debate,[57] the Scots were exempted from paying any duty on Scottish salt for seven years after the Union, and thereafter were to pay only 1*s*. per bushel as opposed to the 3*s*. 4*d*. payable by English consumers.[58] While it is not clear from the available evidence whether the saltmasters were amongst the many supporters of this amendment to the original proposals, it seems hard to believe that they would have missed this opportunity of ensuring that domestic salt should be widely available on advantageous terms within Scotland. This was an important concession for the Scots. It has been suggested by P W J Riley that in the final vote on the Union, which ratified the whole Treaty, there was little correlation between a member's voting record and his ostensible economic interest.[59] This, however, does not exclude the possibility that during the earlier debate, landowners who had an interest in Scottish salt did not do all in their power to ensure its survival.

And as has already been demonstrated, over the course of the next century, it was the domestic market upon which the saltmasters' survival depended. They had, however, to be continually on their guard lest any chinks were allowed to appear in their protective armour. Although no formal records of saltmasters' meetings have survived, it is clear that they did on occasion gather together. There is, for instance, an entry in the cash book which belonged to the factor of Prestongrange, for £6 Scots which was spent in January 1717 at 'ye meeting of Saltmasters in Scotland who met in Edr...to Consert matters about Scots' salt'.[60] In 1733 the 'Proprietors of Salt Works' on the Forth presented a memorial to the customs authorities in which they complained about the 'great decrease of the Sale of Home Salt by Reason of the frequent Consumption of Forreign Salt', and urged that the laws concerning the sale of such salt be strictly upheld.[61] They were not slow to exploit to the full their friends in high places and what has been described as that 'peculiar connection which existed between the Treasury bench and the Scottish patronage system'.[62] Thus, during the 1740s, proposals by Robert Lang and Company that they should be allowed to import rock salt into Scotland were turned down, evidently on the advice of Lord Islay, 3rd Duke of Argyll.[63] It was not an immense organisational task for the saltmasters to marshall their forces to oppose unwanted change. After all, there were only three dozen saltworks on the Forth in the early eighteenth century. At least ten of these, mainly in the Prestonpans precinct, were small and their owners appear to have played no part in any of the saltmasters' agitations. As on some occasions more than one works was owned by a single proprietor, as in the case of Colonel William Erskine, who in 1716 owned pans at Kincardine, Newpans and Torry, it seems that the salt 'interest' was effectively in the hands of less than twenty individuals. Later in the century, meetings of the more formally constituted 'Salt Association' rarely involved the representatives of more than nine or ten saltworks.

Up until the 1770s any threats to the ascendancy of the Forth saltmasters could apparently be countered by an informal gathering, from whence pressure was applied in appropriate political quarters. Early in 1773, however, a situation arose which forced a number of the saltmasters to abandon their formerly casual approach to collective action and create a formal body, the 'Proprietors of Salt Works on the Firth of Forth', or what was more commonly referred to by contemporaries either as the 'Salt Association', 'The Salt Society' or simply 'The Society'.

The reasons for its establishment are not hard to find. The intense boom in Scottish economic activity between 1769 and 1772 was shared by the salt industry. Unusually, salt girnels were completely emptied and advance orders for salt flooded in. On the 1st day of March 1771, for instance, the factor of Methil and Wemyss saltworks wrote to a potential customer that, 'there is not at present any salt to be had in all this Frith, what Triffles may be in the Girnels being all Sold and Vessels taking it away'.[64] In Scotland as a whole between 1768–9 and 1772–3, sales of Scottish salt rose by 21 per cent, to over 318,000 bushels, several thousand more than the previous highest recorded eighteenth century figure. Such intensity of demand and the high prices which accompanied it inspired a round of saltwork construction which was reminiscent of the boom periods of the 1630s and the 1690s and early 1700s. Old and ruined panhouses and pan shells were recommissioned, as at Grange, where five old pans were brought back into use.[65] The most spectacular development of all took place at St Monance, where the first of a battery of nine pans began to make salt in July 1772, just after the boom had broken.[66] At the very least, the net result of all this activity was the addition of some 60,000 bushels of productive capacity to the Scottish salt industry, equivalent to over 25 per cent of the average coastal and retail sales achieved in the three years from 1767 to 1770.[67]

The ensuing slump left many of the industry's leading producers with a stark and unprecedented problem of surplus capacity. Short-term fluctuations had been occurring throughout the eighteenth century, with formerly 'silent' pans being brought back to life during periods of intense demand. At the same time, however, there had been sufficient drift out of the industry to ensure that the balance of supply and demand was kept roughly in equilibrium. In the early 1770s though, the balance was seriously skewed and drastic action was required if potentially fatal cut-throat competition was to be avoided. In addition, after having endured decades of relatively low prices, of on average, around 1*s*. per bushel, before duty, the saltmasters had enjoyed, during 1771 and the first months of 1772, prices which had gone as high as 1*s*. 4*d*. They were reluctant to see them fall back to their previous levels. Hence in the middle of February 1773 moves were afoot, led it seems from Wemyss Castle and with Dr John Roebuck at Bo'ness playing a leading advisory role, to persuade a number of the Forth salt proprietors to act collectively to resolve their difficulties.[68] Data relating to productive capacity, stocks and selling prices was quickly gathered and by the 25th it was being urged by the salt factor at Methil and Wemyss that some salt should be exported, 'to keep up the price and quicken the home sales'.[69] To take matters further a formal meeting of those concerned was called for 10th March, in Edinburgh's Exchange coffee house.[70]

THE 'PROPRIETORS OF SALT WORKS ON THE FIRTH OF FORTH'

Although the Salt Association's books and papers appear not to have survived, miscellaneous material relating to its activities can be drawn from the record collections of some of its individual members, thereby making it possible to describe its activities in some detail. Although not all saltmasters became members of the Society, it was always an association of river Forth producers. Support was strongest from those large works which relied heavily upon coastal sales or markets in the west and where there were relatively few opportunities for landsales. Often such works were those which were linked to collieries which had not yet freed themselves from their former dependence upon salt pans and thus leading lights in the Association were Dr John Roebuck and the Cadells of Grange, as well as representatives of the Wemyss estate and the Newark Coal and Salt Company, at St Monance. On the other hand, there was little active support for the Association from owners and lessees of collieries and adjoining saltworks in the vicinity of Prestonpans. Similarly, further to the west, around Musselburgh, pan proprietors such as Sir Archibald Hope and the Earl of Abercorn declined invitations to join.[71] Their markets were reasonably secure and growing and as with the panmasters at Prestonpans, they were able to benefit from the higher prices which they could charge for landsales. Edinburgh and its hinterland could provide such security for only a limited number of works and at one time or another most other saltmasters on the Forth seem to have been Association members. At no time does the Association appear to have represented less than 55 per cent (or 61) of the 110 pans which were to be found on the Forth in the 1780s.

While some saltmasters saw no benefit in joining, others were simply not asked. The partners of Halbeath colliery and saltworks decided in 1786 not to join, probably for the very good reason that as new entrants to the trade they saw that they could reap the benefits of membership without being burdened by its cost; and as one of their number, Thomas Fishwick, reflected, 'while the above Society can obtain 1/4 per Bushel clear for their Salt they will hardly be so mad as to reduce it to 10*d.* again in opposition to us.'[72] Others left or threatened to leave following disputes about their quotas. At one time Fordel saltworks had been restricted to making 6,000 bushels. As their pans were capable of making much more than this, there was no sense, wrote the salt agent, in keeping 'our pans going for most of the present year and giving away the whole of our profits to the Society's fund which will absolutely be the case if we continue in the association.'[73] On the other hand, although salt from Largo saltworks was being offered at 1*d.* per bushel below the Association's price in October 1774, the works was considered too small to be of any consequence and its harbour too difficult to load from to be of any benefit to an organisation the success of which was partly dependent upon the ability of its members to ship salt out of the river Forth area.[74]

The Association's broad strategy for survival was straightforward. The quantity of salt available for sale on the east coast of Scotland was to be restricted and the price level thereby forced up. This was further adjusted at regular quarterly meetings which also closely monitored members production, sales and stock levels, all of which were governed by the Society's 'Rules'. A mutually agreed output quota was set, at roughly 2,800 bushels per annum for each pan, which was substantially less than the 4,000 or 4,500 bushels which could normally be expected from a well-run, trouble-free pan. The more pans a works had, the greater was the quantity of salt which could be put out. Thus in 1789 allowable production from the Earl of Wemyss' 16 pans was 45,000 bushels, whereas Sir John Henderson, with only four pans at St Davids, was restricted to 10,000 bushels, although this was an increase on what the Fordel works had been allowed formerly. Henderson's allowable output, which was less than 2,800 bushels per pan, was smaller owing to his having two pans which were below the average size, from each of which he was only allowed to put out 2,000 bushels.[75]

Members who exceeded their quota were required to pay a fine or levy of 6*d.* for each surplus bushel into the Association's funds, held on account with the Edinburgh banking firm of William Forbes and Company. Surplus salt was sometimes sold by the Association's secretary, but more often by the members themselves, amongst whom the salt from over-producers was redistributed. It was crucial to the success of the scheme that such salt should not enter the home market, which meant in effect, eastern Scotland. Thus in December 1775, John Grieve of Grange saltworks informed his partners that they had to dispose of 4,705 bushels of Wemyss salt by Whitsunday next, 'and that to Fisherys, England or Exportation, but not to Scotland'.[76] Encouragement to get salt away from the east coast was therefore provided to members, in the form of subsidies: 4*d.* per bushel for salt sent to Dumfries, 8*d.* if it was sent to England or to the Iceland fisheries. To discourage east coast sales as well as to raise funds for the Association, a levy of 2*d.* per bushel was charged on each bushel of salt sold either direct from the pans or which was sent coastwise for sale on the east coast, regardless of whether or not such salt was part of the allowable output.[77]

Overall control of the Association's activities was in the hands of a secretary, who for a considerable time was William Wilson, the salt factor at Methil and Wemyss, based at Wemyss Castle in Fife. Considerable effort was required of him. For the quarterly meetings he had to produce detailed accounts of members' output, sales, stocks and financial position vis-a-vis the Association. Minutes of meetings were carefully kept and circulated. Equally important were his administrative functions. The most important of these was to organise the redistribution of surplus salt and to deal with the sort of request which came from Fordel in October 1789. James Pinkerton, the salt agent there, reported that he was 'overloaded' with salt and that 'relief must therefore be given us, or we must stop our Pans as our Granaries will not hold above two or three weeks more salt.'[78]

In assessing the impact of the Association it is extremely difficult to disentangle the effects of its attempts to restrict output from the natural downturn which was to be expected after the boom ended and a period of less excited trading conditions began. Recorded sales of Scottish salt did fall, from an average of some 313,000 bushels in the three year period 1772–5, to 253,000 bushels in 1787–8, the lowest figure recorded since the mid-1760s.[79] It was not until the period 1795–8, when support for the Association was crumbling, that the average annual sales figures surpassed the levels they had reached in the early-1770s. If aggregate figures only hint that collective action may have held the sales level down, closer investigation justifies a more positive conclusion.

As has been seen, one of the Association's principal aims was to limit the supply of salt in the east by subsidising its sale elsewhere. Recorded sales of salt for fishery use quadrupled between the three year periods 1770–3 and 1780–3, while the share of this market of total Scottish salt sales rose from an average of around three per cent in the late 1760s and early 1770s to over seven per cent during the 1780s.[80] Although opportunies for exports were few and disappearing, the effort that the Association put into this trade does appear to have maintained it beyond the the end of its natural life. As a proportion of total sales, exports, which had fallen to below one per cent by 1771, picked up to average between five and six per cent in the decade after 1773.[81] Unfortunately there is no official record of the destination of coastwise shipments, and anyway the extent to which any increase in these to Galloway can be attributed to the Salt Association is blurred by the fact the Cadells, at Grange, had begun to send salt there in the later 1760s.[82] All that can be said with any degree of confidence is that shipments to Galloway were encouraged by the Association and that large shipments of salt, of over 2,000 bushels, were regularly sent to ports such as Dumfries.[83]

The total quantity of salt disposed of in these ways was relatively small—only occasionally did it rise much above 20,000 bushels, equivalent to around seven per cent of total average sales of Scottish salt in the 1780s. Yet evidently this was sufficient to allow the Association to exercise some control over the price level. Of course market pressures operated too, but it is notable, for instance, that it was not until August 1774, two years after the boom had broken, that 'at a general meeting of the Salt Masters', the price of inland salt was reduced from 1*s.* 4*d.* to 1*s.* 1*d.* per bushel.[84] By the mid-1780s, with members' output evidently well under control and regular channels for the disposal of the surplus having been established, the price of a bushel of salt had been increased to 1*s.* 6*d.*[85] For those proprietors who could keep their production costs down in the region of 9*d.* or 10*d.* per bushel, the gross rate of profit could be around 100 per cent, although the Association's sales tax of 2*d.* a bushel did reduce this somewhat. Ominously, but nevertheless an indication of the grip which the Association had on the salt trade over a large part of Scotland, mutterings began to be heard about the existence of an institution which 'appears to be of the monopolising kind, tending to keep up the price of salt'.[86]

Thus far, attention has been concentrated on the economic functions of the Salt Association. Aware of the hostility which would result if public interest in their anti-social activities was awakened, members did their best to maintain a low profile. In October 1773, for example, it was reported to William Cadell that Society members had decided not to apply to Parliament for legislation which would prevent Irish salt being smuggled into Scotland under pretence of curing fish, as they would be 'exclaimed against for Monopolisers'.[87] From the middle of the 1780s however, the Association and what it represented was being subjected to growing pressure from a number of directions, which ultimately forced its members to adopt a higher profile than many of them felt comfortable about.

As was indicated above, there was increasing concern about the price of salt in Scotland. The prices charged for salt to domestic consumers had always been higher than those which applied to bulk purchasers. Under the Association, however, the differential became wider than ever before, turning quiet resentment into public anger. In 1779, for example, according to price lists set by the Association, 'consumers' were required to pay 2*s.* per bushel of salt, before duty, whereas the saltmasters' surplus was being sold to the 'Fishery' at half that price.[88] Along the north east coast too, louder voices began to be heard about the duties which were payable on both coal and salt. Charges which had first been made in the 1660s by Aberdeen Town Council about the 'improper monopoly' of the saltmasters were revived.[89] From Hopetoun House, John Hopetoun was moved to argue that the duties on coal and salt should be removed. Cheaper coals, he wrote, 'would chear our distant cold Gloomy Coasts', while an alteration to the Salt Laws would 'benefit the Youngest Child of the poorest person who pays for Salt for its porridge.'[90]

For over a century, pleas of this sort had fallen on deaf ears. However, with the ascendancy in Scottish politics of Henry Dundas, MP for Midlothian and since 1775, Lord Advocate, there was now good reason to suppose that arguments such as those developed by Hopetoun would be favourably received. Anxious to extend his influence in the north east of Scotland, Dundas was driven to act in the interests of men whose main objective was to secure 'change in the economy in order to make them richer.'[91] One of the many obstacles Dundas's potential allies in the north east faced was the unnecessarily high price they had to pay for coal and salt. It was the duties on coal, from which shipments of coal within the estuarial limits of the Forth were exempt but which added 3*s*. 8*d*. to the price of a ton of coal shipped from anywhere else, which went first, in 1793.[92] The saltmasters were well aware of the importance of this step. It was, wrote Sir John Henderson of Fordel, then president of the Salt Association, an 'artful prelude' to unfavourable changes in the Salt Laws.[93] Indeed the saltmasters had been active in opposing the 'Coal Bill', not only because by risking the viability of some collieries it could limit the availability of panwood for making salt, but also, in some cases, they judged that it would be 'highly prejudicial to their Interests as Proprietors of Coal'.[94] Significantly, it was one of Dundas's close allies, David Scott of Dunninald, in the county of Angus, who benefited most from the arrival of cheaper coal. Intent on breaking the monopoly of the Forth saltmasters, with lower coal prices and a rise in the price of salt, he was able in 1794 to begin to construct a saltwork on his estate and supply some part of the local market which until then had been entirely at the mercy of the Forth saltmasters.[95]

Around this time too, further discomfort was felt by the Forth saltmasters from events taking place in Ayrshire. John Taylor WS, from Edinburgh, had first developed an interest in coal mining in Ayrshire, at Newton-upon-Ayr, in 1786.[96] Soon afterwards, however, he had turned his attention to the possibility of making refined salt. For such a venture to succeed of course, he had to be able to import rock salt from Liverpool. Hence he began to play a prominent part in a revived agitation which had been started in the mid-1780s by Lord Dundonald to persuade Parliament to allow the free importation of rock salt into Scotland. Taylor's publicly aired case, which was apparently supported by other coal and salt interests on the west coast, was both straightforward and convincing. 'Immense' coal reserves in Ayrshire and Renfrewshire, which could be mined much more cheaply than on the Forth, could serve new pans on the Clyde. These could be supplied relatively cheaply with Liverpool rock salt. Indeed even some of the Forth saltmasters recognised that there was much sense (although little justice) in this line of argument: 'There is no doubt', wrote James Pinkerton, the salt agent at Fordel, in March 1788, 'but the introduction of Rock Salt would be a great advantage to the Country in general by reducing that necessary article of Life', but, he went on, 'it is unreasonable that individuals should suffer by it.'[97] In the light of views such as this the repealers were clearly correct in arguing loudly that if rock salt could be freely imported, the salt industry would 'no longer [be] in the hands of monopolisers [and] we should have it [salt] always in plenty, and at a reasonable price.'[98] Taylor's case evidently hit a tender spot in the east: with unwitting irony John Cadell urged Henderson of Fordel of the urgent need to expose to the public Taylor's 'self-interested views', while remarking that it was 'very curious that by a dash of a pen of a George Street writer', 158 salt pans would be 'annihilated', ten collieries stopped and hundreds of workers thrown out of employment.[99] Unhappily for Cadell, it had all been heard before.

In their final desperate struggle to retain their monopoly of the Scottish market, however, the saltmasters did employ one new argument, which had been used first by the Forth coalmasters (a number of whom were also involved in the Salt Association) in 1793. Like the coalmasters, they appealed to the Act of Union for their defence. Coal shipped within the Forth estuary had been exempt from taxation since 1707. Similarly, the saltmasters claimed, 'The policy of exclusion of English salt has been adhered to ever since the Union', and successive ministers of the crown had upheld 'what has been so long established.'[100] Accordingly, any change now would be a violation of the Treaty. At the very least, urged William Cadell of Grange, it was 'surely Just and Equitable', that the proprietors of saltworks should be indemnified for the losses they would sustain as a result of the proposed changes in the Salt Laws.[101] This was also the view of the James Pinkerton of Fordel colliery and saltworks, who pointed out to his employer Sir John Henderson in 1788 that Fordel colliery did not depend upon salt production for its profitability. Nevertheless, it was argued by Pinkerton, if it had been known that such changes in the Salt Laws were being contemplated, 'we would not have set about erecting any more pans at St Davids'.[102]

This new line of argument was followed in February 1794 by an act of almost breathtaking hypocrisy. Alert to public anxiety about the current shortage and high price of salt, at a salt proprietors' meeting at which were either present or represented, James Christie of Durie, the Cadells of Cockenzie and Grange, Colonel Wemyss of Wemyss, Sir John Anstruther, the Halbeath Company and Colonel Callendar of Crichton, the president, Sir John Henderson of Fordel, read out the Act of the Scottish Parliament of 1573 which had banned the export of salt from Scotland. In an empty gesture which turned the saltmasters' place in Scottish history on its head, the meeting solemnly agreed that they would follow the 'precedent' of 1573 and bind themselves not to export any salt for a year, as long as the selling price in Scotland did not fall below 1*s.* 6*d.* a bushel. Their new-found patriotism and anxiety to serve the inhabitants of Scotland with salt at reasonable prices however, was conditional upon the continued exclusion of English salt from Scotland. In the minute of the meeting, which was to be sent to anyone in Parliament who had an interest in marine salt manufacturing in Scotland, there was also an undertaking to 'Erect or Repair furthwith' as many saltpans as they could.[103]

The Forth saltmasters' pathetic latter-day conversion to the national interest came too late. In 1798, for the first time since 1707, the 'sacred ratio' (3:10) of low Scottish to high English salt duties was altered, with the respective levels of duty being raised to 6*s.* 6*d.* and 10*s.* per bushel.[104] Although John Taylor's bid to get parliamentary approval to import rock salt to his pans at Prestwick failed in the same year, it was not opposition from the Forth which defeated him, but instead the vested interests of English refiners such as Thomas Bridge of Davenham, near Chester, who wanted to restrict the spread of salt refining.[105] The upturn in the market for salt in the second half of the 1790s and the strengthening of the coal industry removed the economic justification for any form of defensive organisation, except amongst less than a handful of the Forth saltmasters.The influx of new entrants to the trade thereafter made it extremely unlikely that effective collaboration could take place.

DEFEAT DELAYED

By the end of the eighteenth century the notion that Scottish salt should be taxed at a different level than the English product was rapidly losing credibility, although within Parliament voices could still be heard pointing to the plight of the poor in Scotland, who were 'compelled to live principally on vegetable food, which required a great deal of salt to stimulate digestion'.[106] It is true that in 1799 the excise duty on Scottish-made salt was reduced from 6*s*. 6*d*. to 4*s*., following pressure from saltpan proprietors such as Robert Reid Cunninghame of Saltcoats.[107] Yet Treasury concern is likely to have been greater about the loss of wartime revenue which the rise in duties had inadvertently caused by giving greater incentives to salt smugglers on the west coast, than about the deleterious effects such policies may have had on saltwork profits.[108] Indeed at this stage the distinctively Scottish strands in the history of salt taxation become inextricably bound up with those which comprise the complex and confused tapestry of the British Salt Laws. Pressures from many parts of Britain were growing for the removal of restrictions on trade and industry, from chemists, bleachers and calico printers, for instance,[109] as well as on behalf of the fishing interest, which had long vented its wrath on duties the legal avoidance of which necessitated cumbersome and time-consuming journeys to collect salt from approved salt stores, as well as unwanted paperwork.[110] The minister for Harris was representative of the views expressed by many of his colleagues in the north western seaboard of Scotland in hoping that the legislature would listen 'to the universal cry of the poor people' about the injustice of the Salt Laws.[111] Smuggling was rife, with the minister of the Small Isles pointing out that the his flock would 'have it [salt] on the easiest terms possible, whatever be the means', and that 'the difficulty of obtaining it...encourages an illicit trade.'[112] To the south also, similar pleas and concerns about the immoral impact of the Salt Laws were to be heard. An additional complaint came from Galloway, where the cattle trade was suffering. This was because rather than using the poor quality Scottish salt (which after 1793 could be difficult to get anyway), and unable to afford the highly taxed English article, instead of slaughtering and salting beef in Scotland, cattle rearers were being forced to drive their animals over the border on the hoof, and consequently incur substantial losses in their weight and value.[113]

No matter how convincing the arguments for substantial alterations in the duties and drawbacks on salt, in the short run they were overshadowed by the state's need to raise revenue to pay for the Napoleonic Wars in the early nineteenth century.[114] With the arrival of peace in 1815 however, it was only a matter of time before the salt duties were totally repealed. In 1818 a Select Committee reported in favour of total abolition. Although the devastating effect that such a measure would have on the Forth salt industry was mentioned by one witness, the Forth proprietors were not directly represented. Anyway, the case for abolition was overwhelming.[115] In the House of Commons, the advantages of post-war reductions in the salt duties were to be seen in the increased consumption of salt and the sharp drop in the number of prosecutions for smuggling. On 5 January 1823, the duties on Scottish salt were removed altogether. The opponents of change had been correct to fear for the future of the Scottish industry. Within two years Scottish marine salt producers began to feel the deleterious effects of the arrival in Scotland of large legal shipments of low-priced rock salt from Liverpool.

5

SALTERS, SERFDOM AND STATUS

THE ISSUES

Traditionally, salters have been seen as members of an occupational group whose status was little different from that of the colliers. With the coal miners, it has been said, the salters 'suffered a degradation without parallel in the history of labour in Scotland.'[1] Brutish and brutalised, it is argued, the salters laboured with the colliers under the yoke of serfdom from 1606, when they were barred from moving to a new employer unless they had obtained a testimonial from their former master which declared that he was agreeable to their release from his service. Salters who ignored the law, and who were discovered with a new master within a year and day of leaving the old, were 'estemit repute and haldin as thieffes', and liable to be 'punischet in their bodyes'.[2] The status of the salter was further reduced, it has been suggested, by the power granted by the Scottish Parliament to masters to 'apprehend all vagabonds and sturdie beggaris', and put them to work in the pans. In 1647, the constraints upon their working lives were made more severe with the passage of an Act which prohibited colliers and salters from observing Yule and other 'superstitious days'.[3] What has been described as the 'coping stone' of this iniquitous series of enactments, came into force in 1701, when the same two categories of worker were specifically excluded from the terms of a Scottish Habeus Corpus act.[4]

As was pointed out in the Introduction, the tendency has been to lump the colliers and salters together, and assume that their experience was identical. Until recently, most commentators have been of the opinion that the circumstances of both were wretched. Reflecting the now far from convincing view that the collier-serfs were a caste or 'race apart',[5] the historian of Culross and Tulliallan reported unfavourably upon the 'wild existence' of the salter 'pariahs' on Preston Island in the early decades of the nineteenth century.[6] The Clows, in their monumental study of the chemical industry, evinced horror that the salters' wages were paid in salt, and at their 'ghastly' working conditions. These had inspired one visitor to Bo'ness in 1765, to write that, 'Nothing ever Exhibited Such an Idea of the Infernall Regions as this Horrid furnace and the Poor Miserable Naked Wretches attending it'. 'The girls Hair', he went on, 'were Snakes'.[7] Their suffering, it has been argued, was possible owing to the fact that their occupation was easy to organise along serf lines, being unskilled and often carried out in geographically isolated locations, supervised under the unremitting gaze of their eagle-eyed grieves.[8]

In spite of the superficiality of many of the studies upon which comments of this sort have been made, they do appear to have some basis in fact. The laws which effectively bound salters to their places of work for life were often enforced, as in November 1636, when John Wilson found himself at the centre of a tug-of-war between two saltpan proprietors who claimed he belonged to them. The Laird of Tulliallan, it was alleged, had sent a boat and 'accomplices' across the Forth to Elphinstone pans, where Wilson was working, and 'put violent hands on said John...and strake him', before he was taken and detained at Tulliallan.[9] In 1705, Wemyss of Bogie initiated a search for and the arrest of Patrick Davidson, one of his salters who had left to work at Maitland pans, near Bo'ness.[10]

Like colliers, salters could find themselves treated as part of the stock of the works, as in 1746, when Sir Robert Henderson of Fordel let 'three Salt panns Salt Girnells Salters and Salters houses' at Inverkeithing to John Wilkie for 300 merks, or in 1750, when the Duke of Argyle, anxious to re-let his saltpans and coal mine near Campbeltown, gave permission to James Kilpatrick to 'search for any Coalliers or Salters already belonging to the...works & to compell them to return thereto'.[11] Life bonds too were signed by some workers, such as James Alexander and Peter Pugston in 1757, who agreed that they and their heirs should serve Sir James Wemyss of Bogie 'in the Station of...Salter, all the days of our Lives'.[12] Freedom from such servitude could be hard to obtain, and when granted, could be partial, with the applicant being liable to be recalled at any time. Thus, when in June 1750, David Thomson's request to leave the service of Prestongrange pans was granted, it was conditional upon his master's right to recall him, 'whenever I shall find it necessary'.[13] Punishment, too, of the salters, could be severe. On hearing of a dispute at his saltworks in 1728, James, 4th Earl of Wemyss, had ordered that one salter, Robert Mitchell, be apprehended and incarcerated in Kinghorn prison, while 'young Miller...the ringleader', was to be put into the 'pitt', an essentially medieval place of imprisonment found and used at a number of saltworks which were within the bounds of a burgh of barony.[14] In 1679, the lessees of the pans of Innerwick and Thornton were granted powers to hold 'ane baylie' [a baillie court], 'for punishing and fynding' delinquents, with liberty to put offenders in stocks and jougs, 'or otherwise doe with them as they please'.[15] For stealing salt and iron from the girnel at Bogie pans, William Tilloch or Tulloch, a salter's assistant, along with John Dick, a collier, were whipped before being transported to Holland.[16]

Yet caution must be exercised when interpreting this catalogue of oppression, and care taken to ensure that such examples are representative and reflect something like the reality of social relations and everyday life as experienced by the bulk of the saltworkers. It is also important to distinguish between conditions which were uniquely the result of serfdom and those which were simply the consequences of the technical and organisational requirements of the salt manufacturing process, which transcended national legal systems. Nor should the conditions of the salters or their attitudes be examined solely within the narrow occupational confines of the salt industry, but rather in the wider context of work and leisure in the period under review. Failure to expand the framework to include these considerations has produced a distorted picture of the Scottish salters' circumstances. For example, implicit in the Clows' pessimistic assessment, is the assumption that the partial nakedness and the dank and bedraggled appearance of the salters at Bo'ness was both a function and indictment of their experience as serfs. Yet, notably, the writer of the source which formed the basis of the evidence upon which this assumption was built, used the example of Dr John Roebuck's large Corbyhall pan. With at least 10 salters, and other workers in attendance, the sight of this vast steam-and smoke-filled panhouse interior must have made a striking impression on visitors unfamiliar with the industry. Yet this pan was uniquely large, and was used for five years at the most. What is more, exhausting heat and the resultant semi-nakedness of the workers was not confined to Scottish salt making, and had nothing to do with serfdom. Such conditions and sights were equally shocking to Victorian factory inspectors of saltworks in Cheshire, where serfdom was unknown.[17]

SKILL AND THE SALTWORK HIERARCHY

Paradoxically, the operational requirements of, and problems associated with, salt manufacturing, did much to dilute the potentially immiserating impact of the labour legislation passed by the Scottish Parliament. As was suggested earlier, successful salt manufacturing depended heavily upon the salters' skills. As most salters appear to have been recruited from within the industry (see below), few documentary traces of anything like formal apprenticeship are available, and thus little can be said about the nature of the skills the salter was expected to attain. However, from Dysart in 1752, an agreement between a new entrant to the trade, Walter Smith, from 'Auchtergevan in Perthshire', and David Beatson, factor of the St Clair estate, has survived, and provides some clues as to what was involved. Smith, who undertook to remain at the works for 14 years, was to be instructed in the 'art of saltmaking'. Although no periods were specified, instruction was to be provided by established salters who would 'Conceal no point nor practique' from Smith, who was to be maintained in clothes and receive two pecks of oatmeal each week until he was 'able to work half ane pan'. Thereafter, as an 'able and Sufficient Salters Servant', Smith was to be paid 'the same fees as any other Salters Servant at Dysart', until he had reached the stage where he was 'Capable...to manage and work ane saltpan'. Only at this point would he earn the designation master salter.[18]

Accordingly, the seizure of beggars and other unfortunates to be herded unwillingly into service as salters does not appear to have been a practical proposition for saltpan proprietors, and no evidence that such persons were so employed has been found. On the other hand, it must be admitted that virtually no data on labour recruitment is available for the crucial period of the industry's great expansion in the late sixteenth and early seventeenth centuries. What is clear, however, is that later, saltwork owners and lessees were fairly selective about who they employed. Thus, it was hoped that a shortage of 'sufficient' salters at Bogie in 1722 could be overcome by finding 'four Sturdy Country fellows that are of honest repute', and who would be 'willing to be taught to make Salt'. Later, in 1730, at the same works, William Tilloch was considered capable of managing two pans, while William Foord, at the West Pan, was deemed suitable only to be 'conjunct Master with Thomas Cusin'.[19] It was their need for skilled workers, familiar with the industry, that drove saltmasters, like their counterparts in coal, to poach labour from fellow employers. While some were prepared to tempt workers away from neighbouring works, others preferred instead to have the water of the river Forth between them and the aggrieved party. During the buoyant years of the later 1720s, for instance, several salters were brought to Kirkcaldy from the south bank of the river. In June 1727, over £8 Scots was spent shipping one of them, Mark Muir, along with his 'familie and plenishing', while in November a further 18*s*. Scots was used to pay for the factor's expenses in going 'over the water for Salters'. Another trip early in 1728 was evidently highly successful, with the Bogie Salt Book revealing that 10*s*. Scots was 'Spent at Salters aggreement and to ther welcom over from Lothian'.[20]

Concern with the quality of salt produced though, is unlikely to have been the main reason why masters carefully sought out established salters. Skill, as well as physical fitness and stamina, was necessary simply to achieve the quota of salt which salters were expected to deliver in return for a certain quantity of coal. Again at Bogie, the tutors of the Wemyss estate, late in the winter of 1720, were becoming increasingly concerned about the arrears of salt due by John Hound, who was 'old and not able to work his pan'.[21] Care was taken with replacements. When in June 1779, George Robertson was dismissed from his pan at Grange for only turning out three draughts of salt weekly, the salter who offered to do better, James Stanners, was only to be engaged after a week's trial to ensure that he could turn out the requisite four draughts.[22]

Equally important to saltpan proprietors was that their salters should be skilled and responsible enough not to damage the pans. As has been seen in Chapter One, some repairs to pans could take weeks and seriously interrupt production. Costs of iron, wood and stone, the tradesmen hired to mend the pans and panhouses, as well as the charges involved in maintaining the idle salters, were borne by the employers. It was therefore in their interest to keep repairs to a minimum. Overheating the pan, or failure to stop for 'paidling' regularly, could lead to the pan bottom being burnt and, subsequently, to the appearance of holes in it. Hence, while the 2nd Earl of Wemyss in 1657 was anxious to obtain dry white salt from his salters, the most serious retribution, punishment in 'body and purse', was to befall any salter who burnt or spoiled his pan 'knowingly'.[23] At Dysart, Walter Smith's contract contained a clause which recompensed the owner at the rate of 'two pennies for each pennies loss', which was caused by Smith's 'negligence or default'.

For these reasons then, like pitmen, the best salters were born and not made, and became familiar with the trade early in life. Salt manufacturing, in common with several occupations, many until comparatively recent times[24], tended to be hereditary, with sons following fathers and other close relatives into the pans. Such relatives were usually males. Females were rarely to be found managing pans, and where this did occur, as in the cases of Margaret Brown and 'widow Greg', at Cockenzie in 1718, it appears that they had taken over their reponsibilities following the death of their husbands. Such arrangements were usually temporary, with a son or other male being drafted in as soon as possible, although it does seem that such females performed a critical function during the transitional period, not only keeping the pan going, but probably also completing the training of their replacements. Thus, at Bogie, in October 1729, James Rae was appointed 'master of his father's pan...in place of his mother he oblidging himselfe to maintain her'.[25]

The use of the term 'master' salter is significant. Saltwork employees were not a homogeneous group, a single stratum of workers easily distinguished from their grieves and employers, the saltmasters. True, at the apex of the organisational structure of the saltwork was the owner or lessee, beneath whom was usually a salt grieve, agent or factor, paid at a fixed weekly or monthly rate in return for managing and co-ordinating the affairs of the works. It was a complex task and involved carrying jobs which in the twentieth century would be separately designated, as security officer, book-keeper, clerk of works, buildings officer, production manager, pay clerk, personnel officer, sales manager and executive.[26] Usually these functions were performed by a single individual, although at the larger works two persons were often employed.

Subordinate to the grieve but central to the succesful conduct of the whole operation, was the master salter, usually described simply as a 'salter', but who in the more carefully kept account books would be accorded his higher status. It was his responsibility to operate a saltpan (occasionally more than one) and make salt. Either individually or in combination with others of his rank, the master salter agreed with the saltmaster or his agent to make a certain quantity of salt in return for a fixed quantity of coal. The ratios could alter, with the seasons, with variations in the quality of coal, or according to the size of the pan. Thus at Bogie in 1731, for example, when the master salters' contracts were being amended, it was agreed that there should be two exceptions to the norm: the 'master of the little pan' was to return a boll of salt less weekly than the others for the coal he was provided with, while the 'Master of the Big pan', was to return an additional boll.[27] The core of the contractual arrangement remained much the same from the sixteenth century to the nineteenth. In 1659, for instance, the Earl of Wemyss agreed with his salters that for every 12 loads or 'dozens' of coal set down at the pans, they should each give him in return two bolls and two pecks of salt.[28] Some eight decades later, in 1730, the salters at Bogie were contracted to deliver three bolls of salt for 'each first dozen panwood'. In June 1801, the master salters employed by William Cadell and Co followed long established practice by agreeing to deliver fifteen bushels of salt in return for 10 or 11 carts of coal in the summer and winter months respectively.[29]

Ranking beneath the master salters, indeed often employed by them, were the salters' assistants or 'servants', whose labour was essential in order to carry on the round-the-clock salt manufacturing process. Their assistance was required to keep the fires going, constantly and evenly, in clearing out ashes from underneath the pans, raising water into them, stirring the brine and drawing off the salt when a full pan had been made. They also relieved the master salter during the dead of night. Although the evidence is limited, there are reasonable grounds for believing that there were considerable regional variations in the social composition of the salt working unit and hence in the relationship between the master salter and his assistants. On the south side of the Forth, and particularly at the smaller saltworks in the vicinity of Prestonpans, the family working unit appears to have been usual, with master salters depending upon their wives and children to assist them. Thus, in a survey of the stock and condition of Prestongrange colliery and saltworks in 1748, the ages of several of the master salters, their wives and offspring were recorded. John Thomson, younger, had three sons, aged 20, 17 and 15; James Fleuchar was only 28, with a wife of 30, and was assisted by a James Fleuchar, younger, who was 18. Ninian Moodie, 41, and his wife Margaret, had two sons and daughters, who were 'all very young'. Attention too was paid to whether or not they still lived nearby.[30] No mention was made of non-family members. Further confirmation of this concentration of family work units in the area comes from nineteenth century descriptions of works around Bo'ness, such as the pan at Grange which was worked by 'Jock', whose sole assistant was his daughter, who, with 'purple paws', helped both to make salt and repair the pans.[31] In Ayrshire too, master salters appear to have drawn upon the labour of their immediate families. Indeed, in Ayrshire, it was usual for salters and their families to be accommodated in rooms either adjacent to or above the pans themselves. Hence the appearance of wooden and iron bedsteads in the inventories of west coast saltworks.[32]

Evidence of what to modern eyes may seem seem acutely claustrophobic living arrangements does not appear in east coast records. Instead, rows of salters dwellings, separate from the panhouses, can be identified in several locations, as on the shore at Culross in the 1790s, where there were two ranges, one comprising single storey houses, eight feet high, the other of larger dwellings, 18 feet high.[33] Just why Ayrshire's salters should have been housed so differently is by no means clear, although for saltpan proprietors there were undoubted advantages in having their workers in such close proximity to the pans. Not the least of these was that the costs of building and maintaining separate dwellings for the salters were avoided. There may also have been less likelihood for the salters to merge with the local community, thereby making it more difficult for them to attempt to establish a local black market in salt illicitly removed from the pans. Perhaps the later expansion of the industry on the Clyde and the widespread dependence of masters upon salters from the east made it easier for them to house saltworkers in this way. At some pans, the 'east coast men', appear to have been temporary migrants, returning home after a few months work.[34] Housing designed to suit these circumstances, may, for some salters and their families, have become permanent.

Family employment was markedly less common at the larger multi-pan works which were concentrated along the Fife coast as well as at places such as Cockenzie, on the Lothian shore. Sometimes, it is impossible to say what the relationship between master salters and their assistants was, as at Thirlestane and Corbyhall pans at Bo'ness in 1770, which were then leased by Dr John Roebuck. Of the 75 people employed at the two sets of pans 21, or 28 per cent, were listed simply as 'women servants', while the 41 male saltworkers were listed together as 'Salters & Servants'.[35] At most other works of this size, however, the salters' assistants were generally unrelated to the master salters, who paid their wages. Some masters favoured working with only one assistant; others, as at Bogie and Methil, employed child labour too, with boys rather than girls being preferred. In fact, at Bogie in 1720, it had been hoped that salters' daughters could be persuaded to become coal bearers.[36] Responsibility for searching for and recruiting salter assistants generally lay with the saltwork proprietors. They paid the assistants' annual binding monies or other customary lump sums and incidental rewards. On occasion the master salters were left to find their own assistants however, as in 1726 when John Foord was sent from Kirkcaldy to Cockenzie and Grange, 'in order to procure a new Servant for himself'. It was, however, the master salters who were responsible both for paying their assistants' weekly or fortnightly wages and for their conduct in and around the pans. Thus, in 1732, one master salter in Fife had his 'plenishing' confiscated by his grieve on the grounds that his 'hind salt', or arrears, had been exceptionally great, even though he had protested that this was due to the, 'unfaithfulness of my Servant', who, 'manie a night when it was his turn to work...hurt the pan [and] neglected his work and imployed him self in breaking oppen...girnels', from which he had allegedly stolen salt.[37]

Any assessment of the nature and impact of serfdom in the Scottish salt industry then has to be tempered by the fact that it was organised on sub-contracting lines. The saltmasters and their agents were generally only indirectly involved in the production process, whilst the master salter was the employer and supervisor of labour within and around the panhouses. This form of contractual arrangement was not unique to saltmaking in Scotland, but was also found in Cheshire, the heartland of British salt production, where it was the 'original basis of employment' and survived long into the Victorian era.[38] In Scotland too,it probably pre-dated serfdom and certainly outlasted it, being formalised in nineteenth century account books in which the master salters names were replaced by those of small partnerships or companies, such as Thomas Forrest and Son, James Miller and Co and Alexander Greig and Co, all of whom made salt at Pinkie in the 1840s.[39]

Indeed, if there was a particularly miserable and abused group of workers in the salt industry in the sixteenth, seventeenth and eighteenth centuries, and perhaps beyond, it was most likely to have been the salter assistants. It was they rather than the master salters who were inclined to desert the works and bring salt making to a halt. At Bogie, for example, in the year from October 1724, over eight per cent of significant stoppages were the result of absences by the assistant salters.[40] Few works were spared the inconvenience of their departure, occasionally to work at the 'keels' at Newcastle, but more often as they sought employment such as 'country service', or moved on to other saltworks in the hope of driving a better bargain. All saltmasters at one time or another had to face the trouble and expense of searching for and recruiting replacements. Although labour shortages had largely been overcome by the middle of the eighteenth century, specific local bottlenecks could appear. Thus in June 1780, in order to encourage salter 'servants' to work at Grange, the partners took the unusual step of intervening directly in the master salter-assistant salter relationship and offering to pay a bonus of 4*d.* for every draught of salt they helped to make.[41] The reluctance of many salter assistants to remain at the pans however, was generally due less to the harshness or brutality of the capitalist panmasters and rather more to the ill treatment they received at the hands of members of their own subtly stratified class.

Other onerous tasks, arguably productive of misery equal to that experienced by the master salter serfs, were also carried on in and around the saltpans. One such example was that of the salt officers, men appointed from 1713, when English salt duties first became payable on salt made in Scotland. It was their job to weigh all salt as it was taken from the pans and ensure that the appropriate duties were paid when it was removed for sale from the girnels.[42] Although properly the subject of another volume, it is worth noting here that in return for regular if modest salaries,[43] whatever their rank, local officers were required to work long hours—through the night if necessary and certainly turning up for duty before sunrise on Mondays in order to observe the salters' activities as they stoked their fires—in wretched conditions.[44] Nagged incessantly by their superiors in the Salt Office in Edinburgh to perform their duties ever more efficiently and stamp out the endemic embezzlement of salt from the pans, local officers had frequently to beg for the smallest concessions to comfort. Although salt had begun to be made at the first of nine new pans at St Monance at the beginning of July 1772, and officers had been stationed there from June, it was not until the end of August that permission was granted for them to have a hut built to shelter them at night and during inclement weather. Grudging permission was given for the construction of a building, 'on as frugall a plan as the Case will Admitt', but as late as January of the following year an officer and three watchmen were driven to petition the Commissioners for an annual allowance for coal to heat its 10 square feet of space.[45]

Infirm and retired salters could reasonably hope that their masters' paternalistic instincts would induce them to make some charitable contribution to their maintenance. As late as November 1797, for instance, 6*d.* a week was being disbursed from Pinkie saltworks for 'John Millers Charity'. Salt oficers, however, could be less fortunate, as in the desperately sad case of John How, 'late of the Salt Duty', who appealed in 1738 to the

General Assembly of the Church of Scotland for relief on the grounds that, 'by great Colds and Night Watching', he had contracted 'a Distemper in his Eyes by which he is almost blind'.[46] Seventy years of age, but apparently unable to persuade any minister to sponsor him, his claim was dismissed.

Saltworks also attracted much casual labour, hired temporarily to perform specific and usually mundane or dirty tasks. Aged women, for instance, were frequently to be found washing salt 'pocks', the coarse cloth bags in which salt was carried from the pans to the girnels and thence to waiting ships or other customers. The female 'salt bearers' at Winton were also employed on this task in the early eighteenth century.[47] Later, in the nineteenth century, wooden boxes were used for the same purpose. At Winton in the early 1700s, the women concerned were paid 2*d.* Scots for each bag they cleaned. Disposing of the vast quantities of ash which were produced by the pan furnaces also created regular, if temporary, work. In July 1817 for instance, James Currie was paid 8*s.* for the eight hours he spent clearing away ashes from Dysart pans.[48] Earlier, it was usual at most works to hire 'barrowmen' or day labourers, as required, at the rates which varied between 5*s.* and 6*s* .8*d.* Scots per day in the mid-eighteenth century. Whilst for many such workers their appearance in saltwork account books represents only a fragment of their working lives, some were evidently heavily dependent for their livelihood upon the scraps of work the salt grieve could throw their way. This was clearly the case where accommodation, however humble, was provided. Betty Adamson, for example, had her house rent paid for her by Dysart saltworks in the early nineteenth century. In return, she was expected, when called upon, to wash salt pocks, and when necessary, when Dysart stocks were low for instance, to visit neighbouring works to borrow others. An unnamed woman, presumably of similar status, was also on the books of Preston Island saltworks.[49]

An altogether different category of temporary worker found at the pans was the craftsman or tradesman, skilled artisans who were hired to construct and repair saltpans, panhouses, girnels and the salters' houses. They were generally paid by the day, at established rates, with the completion of their tasks being celebrated by saltwork-funded drink, as at Winton where 16*s.* Scots was regularly disbursed by the grieve in the early eighteenth century, for 'pan roof ale'. For a number of craftsmen, the employment generated by the larger saltworks would have provided a major proportion of their earned incomes in a period when underemployment and irregular working patterns were commonplace. At one typical Fife saltwork in 1715, for instance, between February and August a total of 10 smiths were employed at pan 'beiting'. Of these men, five worked at the pans for an average of just over 10 days each month, while in some months smiths such as David Spence and Robert Logan obtained as many as 16 days work.[50] Although smiths were the tradesmen most in evidence at saltworks, other crafts were also in demand. As was seen in Chapter One, masons were necessary, to rebuild crumbling panhouse walls and fire-damaged chimneys. The services of wrights, roofers, slaters and thatchers were also called for, on tasks which could often take several days to complete. In their turn, such tasks provided more casual work. One such beneficiary was John Boyle at Winton, who, in October 1718, was paid £5 Scots, 'for Loding two thousand and Six hundred divots from ye Moore of Tranent...for Salter houses'. At the same works, tenants from the neighbouring Grange estate were fairly frequently called upon to provide hearthstones for the pans, at £11 Scots each.[51] At least until the early decades of the eighteenth century, local tailors could be called in for anything from half a day's work to a week, making and repairing salt pocks. Paid 9*s*. 8*d*. Scots per day, compared to the lowest smith's rate of 12*s*. Scots, theirs was the least remunerative of the trades. Later, it became common for grieves to purchase thread and cloth and to employ unskilled casual workers to sew the bags.

The construction of new pans and panhouses and ancillary buildings called for the labour and skills of an even wider range of workers. When a salt girnel was being built on Arran in 1714, 70 days work was created for quarriers, while two carters spent at least 66 days carting the stones from the quarry to the building site. Boatmen, limeburners and plasterers also found themselves in great demand, as did Colin Mcalaster and three other colleagues, who each spent 17 days, at a daily rate of 6*s*. 8*d*. Scots, pulling heather for the girnel roof. Other general labouring jobs involved carrying shingle, flagstones and sand from the shore, for the floor of the girnel.[52]

The last, but not the least significant of the identifiable groups of workers employed at the saltworks, were the 'leaders' or carriers of fuel from the coalhill to the pans. As with the others, their numbers varied with the size of the works, but generally each saltpan was served by one leader. Occasionally though, coal could be brought to a pan by two leaders; even more rarely one leader might provide coal for two pans. Whatever their number, at the larger enterprises, they formed a steady stream of traffic between the coalhill and the pans: in the year from 2 May 1690, for example, on average, just over 658 cartloads of coal a month were deposited at Corbyhall pans. Not unexpectedly, the busiest period of the year was from March until late October. Only 43 cartloads were delivered between 21 December 1690 and 4 January 1691.[53] At least until the middle of the eighteenth century, the leaders appear to have been drawn from the ranks of the poorest tenants, and cottars, of the estates on or around which the pans were located. Indeed, tenants on estates such as that which belonged to the Dukes of Hamilton at Kinneil, in the seventeenth century, were obliged and could be compelled to carry coal to the pans.[54] The leaders were usually male, although of the 14 individuals listed at Winton in July 1719, two were female, Margaret Seton and 'Widow Thomson'. As the century wore on however, female employment all but disappeared. Pancoal, which had previously been carried on the backs of human beings and animals as well as in carts, was almost all taken by wheeled carts which belonged to independent farmer carters, into whose hands the local coal carrying trade had largely fallen by the 1770s.[55] At a very few favoured locations, coal was later carried to the pans on waggonways. Indeed, from its opening in 1722, the line from Tranent-Cockenzie was used to take small coal to the pans, although its main inspiration had been to connect the inland pits on the

Winton estate with the coast and ships which would carry 'great' coal to London.[56]

Earlier, the leaders' relationship with the saltwork proprietor is best represented as having been mid-way between that of the free tradesmen and the bound salters. With the latter, they were usually paid a bounty or 'earnest money' at the turn of the year, although this was always less than was received by the salters. At Cockenzie in 1716, for example, the leaders were each paid 14*s.* Scots, the master salters, 29*s.* Scots. Although the rates for carrying coal varied across the country, the general rule was that the leaders were paid according to how much coal they carried. The rates paid also depended on the distance the coal supply was from the pans. At Winton, for example, in 1716, the rate for carrying pancoal to the ten 'wester pans' was 2*s.* 2*d.* Scots per boll: coal taken to the two 'easter pans' paid 2*s.* 4*d.* Scots. For taking coal further, to Prestonpans, the rate was 3*s.* Scots per boll. At most works, on carrying a certain quantity, called the 'evenwood', the leaders were entitled to an additional payment. This usually took the form of a quantity of free coal, often termed a 'maggwood'. At Winton, two 'maggwood bolls' were paid after 80 bolls of panwood had been taken to the pans at Prestonpans; at Bogie each leader had to carry only 30 bolls in order to obtain a load of great coal, which was either burned in his or her own hearth, or sold.[57] On the other hand, although they differed from most salters in being free to attend to their agricultural duties when necessary, saltwork proprietors evidently felt no obligation to support them during times of hardship, when the pans were 'silent' for example. Some worsening of the circumstances of the leaders who had been drawn from the ranks of the lesser tenants can be discerned on Kinglass estate in the early eighteenth century. Formerly, they claimed in a petition they sent to the Duchess of Hamilton in 1716, they had been 'very kindlie...and fairlie...compted and cleared with as to our Services in our Leading of pan-wood from the coal-hill...to the Saltpans at Thirlestane'. More recently, they complained, their houses, which had formerly been repaired for them, had become 'ruinous'. To make matters worse, they had insufficient work to enable them to pay their rents.

The saltwork then was a complex social organisation in which the master salter had a highly respectable role. Even their assistants did not occupy the lowest rungs of the occupational hierarchy. It could, of course, be argued that this did not preclude the salters from being singled out for especially brutal treatment, backed as their masters were by a powerful armoury of regulatory legislation. And indeed there were many instances where salters were subjected to what, judged by modern standards, would be considered severe punishments for workplace related failings and misdemeanours. As has been seen, corporal punishments were not unknown. More common was imprisonment, although this was usually only for a period of days. Incarceration could follow a strike, as it did in July 1719, when the salters at Bogie had 'set the pans', complaining that the poor quality of the coal they were being supplied with made it impossible for them to make the minimum quantity of salt required by their contracts. For this 'fault', the factor imprisoned five of their number, three in Kirkcaldy jail and two in Aberdour.[58] To fall too deeply into debt by failing to deliver the contracted amount of salt for a given quantity of coal, could also lead to a short corrective stay in jail. This form of punishment, however, was most frequently reserved for salters who were too frequently caught stealing salt from either the pans or the girnels, and selling it illegally, that is, after 1713, without the knowledge of the salt officers. Few saltworks for which records have survived went for more than a year without one or more salters being imprisoned for this reason. One of the most striking examples comes from Wemyss in March 1753, when the saltworks was virtually brought to a standstill as five salters were jailed in Kirkcaldy until they had paid their fines, which totalled £108 Scots, for 'clandestinely selling of Salt to salt Cadgers.'[59]

Yet, in the contexts of seventeenth and eighteenth century Scottish society and the emergent industrial economy, the punishments meted out to the salters were fairly unremarkable, typical rather than unique. Barony courts could hand out similar retribution to anyone who came under their jurisdiction, at least until 1747, when coal and saltmasters were exempted from the act which abolished heritable jurisdictions.[60] There is no evidence to suggest that their continued existence had any marked adverse effect on the salter communities. Authority in early modern Scotland was sternly exercised in many situations. It was in 1700, for example, some time before any salters set foot on the island of Arran, that the much admired, estate-improving Anne, Duchess of Hamilton, ordered her chamberlain to 'furnich a man for Scourgeing and punishing Malefactors'.[61] It was in the 1680s that the proprietors of the woollen mill at Newmills decided that the works could be better managed if part of a stable was converted for use as a prison and that workers who 'broke their engagements', should be brought back forcibly.[62] Indeed, in common with workers elsewhere in Britain who were employed in the rising manufacturing sector, it was above all with threats of dismissal and fines that the salters were disciplined.[63] Decisions to enforce these measures may sometimes seem arbitrary and unnecessarily harsh, as at Bogie in 1725, when the tutors of the estate determined to replace the brothers Andrew and Patrick Hounds on the grounds that they were 'continually quarrelling' (although Andrew was also judged to be a 'lazy fellow') and agreed to retain them only after they bound themselves to behave, on pain of a massive penalty of £20. The strategy appears to have been succesful, the factor reporting some months later that 'they now pay up their Taught pretty well'.[64] If the threatened fine was large however, it was at least as understandable as some fines imposed at cotton mills later in the eighteenth century, where even an ill-advised stare by the unfortunate possessor of an ugly face could apparently incur a financial imposition.[65] In both cases, the employers' use of this sanction could be justified by their desire to ensure that the then detested regime of continuous employment[66] and regular output was maintained.

SOCIAL ATTITUDES AND LABOUR RELATIONS

The concept of 'paternalism' has often been used to depict the nature of social and labour relations in Britain for much of the period with which this book is concerned. Once defined as 'the maintenance by the employer of a strict fatherly discipline...combined with a measure of genuine benevolence',[67] it has been criticised as being too blunt an analytical tool to apply to master-employee relations in the kindred Scottish coal industry.[68] It generally represents a view of society as seen 'from above' and tends to discount or obscure the existence of what has been described as a 'distinct plebeian culture'.[69] This can be defined in part as an alternative perception of the way society and work should be ordered and could be a very real constraining force, the most public expression of which may have been the riot,[70] upon the actions of the 'patricians', commonly represented in the case of the salt industry in Scotland by members of the exceptionally powerful Scottish landed class.[71]

Blunt the tool may be, but the notion of paternalism in industrial relations in salt manufacturing cannot be disregarded altogether. Such sentiment, for example, is prominently reflected in the language used by David, 2nd Earl of Wemyss. In a diary entry for October 1675, he noted, evidently with some satisfaction, that he paid his men well, at rates, 'that none in Fife give their salters'. This, he hoped, would, 'make me in a condition with my salters as any in Fife are.' At the same time, as has been pointed out, he was prepared to take firm action against those who performed badly at the 'sufficient good well-repaired pans', he believed he had provided.[72] Apparently spontaneous acts of proprietorial charity were also commonplace, with sick salters being given some financial support, while those who had died were often provided with a coffin and the living with ample quantities of drink with which to ease the pain of their colleagues' passing.[73]

Nevertheless, such displays of evidently unsolicited kindness can equally be interpreted as essentially self-interested actions, designed not simply to win the gratitude and esteem of workers, but also to command their loyalty to a particular master and set of pans. For, in spite of appearances, dependence was by no means the burden of the salters alone. The 'stigma' which many argue appears to have attached to bound labour, the unpopularity of regular, repetitive and physically arduous work, as well as the skills required of the master salters, meant that the latter were in relatively short supply. Within the confines of the estate saltwork, this gave them a status and authority, as well as bargaining power, far beyond that to which they were entitled by law. Masters then, in spite of the resentment which some felt about their vulnerability to the potentially harmful actions of aggrieved saltworkers, were obliged to treat their social inferiors in a far more circumspect manner than they would have liked. Careful judgement had to be exercised by masters in ascertaining what orders could be given arbitrarily and when their workers' complaints were legitimate and should be heard.[74]

Indeed, close scrutiny of the available evidence reveals that the widely held view of the salters as lowly immiserated prisoners of their saltpans is highly misleading. It is notable, for instance, that in spite of the paternalistic tone of many of the entries in his diary, the Earl of Wemyss was engaged in frequent negotiations with his salters over their contracts. These were agreed upon after discussion and not simply imposed by Wemyss, who was also bound by their terms. Thus, he noted in May 1659, after a new agreement had been reached with his salters, 'I must give them a firlot of meal if they [the pans] stand still in storms' and 'uphold all other things about them, as buckets, stone, lime, shells, leather and the like'. Failure on the part of a master to keep his side of the bargain could result in the salters taking collective action and striking. In 1726, for example, the salters on the Winton estate, then in the hands of the English York Buildings Company, refused to work for more than two months, following an unspecified breach of their agreement on the part of the company's local agent.[75] Salters too demanded that they should be paid regularly. In March 1689, at Cockenzie, it was considered 'a necessitie' that the salters be paid what was owing to them, for otherwise it would 'occasione to cause ane mutinie amongst ym'.[76] The salters did not always have to depend upon the strength afforded by collective action to make their point. In July 1711, Daniel Kerr, a salter employed at the single pan at the Cock of Arran, simply refused to 'cast', or make, any salt until a new agreement had been reached. He was replaced, but appears to have suffered no other retribution for taking the stand he did.[77] Individual action could, on occasion, produce beneficial results for salters. In May 1725, David Tilloch, frustrated by the intrusion by fishermen at Kirkcaldy who were using his bucket pot in which to store live bait, 'set his pan' and refused to 'lay her to' until his bucket pot was cleared. Within days his factor had acted to remove the offending material.[78]

Confrontations of this sort however, were not regular occurrences and disputes which led to strikes appear to have been less widespread than was the case in the coal mining industry. Disputes such as those just described tended to take place when a saltmaster had strayed beyond the bounds of what the salters deemed to be legitimate behaviour. For their saltworks to succeed saltmasters were to a large extent dependent upon the willing co-operation and sometimes the guidance of their master salters. Thus, the Duchess of Hamilton's decision to establish a saltwork on the island of Arran in the early eighteenth century, was only taken finally after a small party of salters from the family's pans at Bo'ness had visited the west coast site, inspected the coal and tested its suitability for making salt in pans belonging to the Cunninghames of Auchenharvie at Saltcoats.[79] As far as can be ascertained, they, along with several other colliers and salters from Bo'ness who followed them to work on the island for periods of varying length, travelled freely, on hired horses, and were paid travelling expenses which included sums for food, drink and overnight accommodation. In an effort to increase efficiency at Bogie, in January 1727 John Foord, described by the factor as the 'lad of best Sense among our Salters', was sent, on his own, to Cockenzie, in order to learn something of the superior salt making techniques used there, in the hope that they might be introduced at Kirkcaldy.[80]

Potential tensions were relieved by the judicious provision of liberal quantities of ale, occasions which were far from the random events they might at first sight appear to have been. By providing his salters and their assistants with £12 Scots worth of ale at the turn of 1718–9, William Adam was not simply making a gesture of seasonal goodwill, but adhering to long-established custom. To have disregarded it would have been been an act of considerable folly, not least because without drink the salters would have been far more likely to have quibbled over the terms of their new agreement, or complained about the amount of their 'earnest money'. Ale, with its power to dull the critical faculties, was also much in evidence on the introduction of new 'firlots', the wooden buckets or containers used to measure how much salt the salters delivered to their masters' girnels. The more salt they were capable of holding, the greater was the advantage to the master. The less fit the salters were to inspect the firlots carefully, the better.[81]

At least until the later stages of the eighteenth century, saltmasters, in common with other manufacturing employers in the early stages of industrial capitalism, were obliged to respect their workers' affinity with a culture in which work was evidently perceived to be part of a lifestyle which included time for feasts, fairs and holidays.[82] Nowhere can this be seen more clearly than in the account books from Cockenzie saltworks in the early 1700s. These reveal that saltwork funds were used to buy coal for the bonfires and saddles for the races which accompanied Cockenzie fair. Neither was colour or spectacle overlooked: in October 1718 some ships' boys were paid £1. 14*s*. Scots to hoist 'their Colours on the Said fair day Conform to Custom'. Several payments were made to pipers and others for 'proclaiming the saddle' in surrounding towns such as Dalkeith and Haddington. A similar role was played by Dysart saltworks in the local 'Gallowtun Race'. In May 1719, as far as can be judged, all of the workers at Cockenzie pans participated in the King's birthday clebrations, warmed by a bonfire and with conviviality ensured by the contents of two barrels of ale and half a dozen bottles of claret. Similarly 'Crisenmass' was better enjoyed by the salters who worked for Carron Company in 1759, who provided them with a gift of 'free Ale'.[83] Evidence such as this must cast serious doubt on the proposition that salters were a 'race apart'. They clearly shared many of the values of other ordinary Lowland Scots. Cockenzie saltworks, for instance, including the salters, played an integral part in the social life and festive calendar not only of Cockenzie itself, but also in some of the inland towns and villages of East Lothian.[84]

Changes in traditional attitudes to work and leisure and the inculcation of regular habits amongst the salter population, were, by and large, considered more likely to be achieved by offering cash incentives rather than by the use of more violent methods of persuasion. Despite the legislation of 1647 which had outlawed such celebrations, around Yuletide, New Year and Handsel Monday for example, the salters were often tempted to take an extended break, which could, on rare occasions, stretch to several weeks. Hence it was usual to pay a bonus to the salter who made the first panful of salt, as in the case of the Cockenzie master salter, James Brown, who was paid an additional 12*s*. Scots in January 1719, 'for bringing in the first Salt this year'. Much later, in 1777, George Stirling, a master salter at Methil, was awarded an extra allowance of pan coals, with which he could make additional salt, because he had begun to work his pan on the 13th of January, 'in opposition to the threatnings of his neighbours who cast theirs idle all that week'.[85] Clearly a man who was prepared to disregard informal collective decisions by the salters to restrict their output, Stirling was also rewarded for his 'extra delivery of oversalt beyond the other salters during the year'. Similar incentives were introduced at Grange in 1779, but rather than isolating and rewarding individually impressive performers, the proprietors tried to utilise to their advantage the salters' tendency to act collectively to control the output from their pans. It was agreed with the salters as a body that if, between them, they made 16,000 bushels, they would be paid a premium of £5: on the other hand, if the salters failed to achieve the target, they were obliged to pay £5 to the company.[86]

EARNINGS AND STATUS

Thus far little attention has been paid to the question of salters' earnings and incomes. This, however, is an important subject, which contributes much to our understanding of the life experience of saltworkers. Wage rates and income levels are also valuable if imperfect indicators of occupational status. Were the salters able, like the colliers, to turn serfdom on its head, and command wages which are often thought to have been significantly higher than those of other members of the Scottish labouring classes?[87] The indications are that they may have done so, yet for a variety of reasons, even after a period of extensive research, it seems that such a conclusion must for the present remain tentative. To attempt to determine with any degree of certainty the income levels of Scottish saltworkers during the period under review and especially up to the end of the eighteenth century, is, methodologically, an exceedingly difficult exercise. Having said this, much potentially illuminating data relating to this topic has survived, and can be used cautiously to add substantially to what little is currently known about this group of workers.

Undoubtedly, the major obstacle to the presentation of meaningful tabulations of salters' earnings is the fact that a significant proportion of their incomes was 'invisible' and went unrecorded by contemporaries. What such sums were, therefore, cannot be known by the historian. What is more, this proportion changed, probably markedly, over time. From the late sixteenth century, and no doubt beforehand, until the early eighteenth century, by far the greatest part of the salters' incomes came from their 'overplus', the salt they were able to make and sell privately after they had delivered to the saltpan proprietor the amount of salt they had contracted to make in return for a given quantity of coal. The nature of these contracts was outlined earlier in this chapter. As Thomas Tucker discovered in 1656 however, it was almost impossible to ascertain what 'overplus' the salters 'truely made', much of it being 'vended and disposed of in the night, or otherwise privatly, in small parcels, without entry', to 'cageors and other poor people'.[88]

That money wages appear to have played little part in the saltmaster-salter relationship for a great part of the period covered in this book, has tended to confirm the generally pessimistic view of the salters' situation.[89] Yet neither poverty nor lowly social status, or indeed an exploitive employer-employee relationship, are necesarily indicated by the apparent absence of money wages, or by the salters' heavy dependence for their livelihoods on the sale of the commodity which they produced. It has been argued that in England in the sixteenth and seventeenth centuries, 'men fought desperately to avoid the abyss of wage labour.'[90] Full-time wage earners were regarded as the poorest members of society, their dependence leaving them exposed to the abuse of the truck system, and the poverty and indebtedness which resulted from frequent lengthy delays in money payments. It has recently been shown that as late as 1850, Dorset quarrymen insisted that they were paid largely in stone, a practice which 'served to deny any real status difference between employer and employee.'[91]

It seems entirely reasonable to suppose that a similar attitude may have prevailed amongst the Scottish salters. As has been seen, the master salters were not simply early proletarians but sub-contractors and small-scale employers. Salt was sought by virtually every inhabitant of Scotland, and until the mid-seventeenth century, with the energies of the saltmasters being directed primarily to overseas markets, the domestic trade was to a large extent the sole province of the salters. Both of these factors combined to give them an independence, as well as a certain status, authority and power in the marketplace, which they would not willingly surrender. As the saltmasters turned their attentions to the domestic trade however, the size and destination of the salters' 'overplus' came under increasing scrutiny. Further tension was added in 1713, when duty first became payable on domestic salt. This was heightened in 1735, when, apparently following pressures from enraged Customs Commissioners whose officers were unable to trace and tax the quantities of salt sold by the salters, Parliament approved a measure which made illegal the payment of salters' wages in salt.[92] The importance of 1735 itself should not be overstressed however: a number of saltmasters, notably on the Fife coast, had begun to pay the greatest part of their salters' wages in cash beforehand, although proprietors of temporarily struggling works, as at Bogie in the 1720s, where cash was in desperately short supply, preferred to maintain the old method for as long as they could. No written record of the salters' reaction to this change exists. What is clear, however, is that they continued, by a variety of illegal means, to obtain salt and sell it from the pans. In this way they kept a foothold in the marketplace and maintained some of the dignity and self-respect felt by workers who were both makers and sellers of the product of their own labour.

The extent to which salters were engaged in the sale of embezzled salt provides the second major obstacle to the presentation of anywhere near accurate estimates of salters' earnings over time. Although the practice was far more frequently noticed and complained of in the eighteenth century, in 1656 Thomas Tucker had observed the salters' apparent inability or unwillingness to distinguish between the salt which belonged to their masters and their own. Poor accounting practices, he held, in addition to the 'vileness and unworthinesse' of the salters, were the principal causes of the problem. Like the size of the 'overplus', the quantities involved could not easily be estimated.[93]

What led the salters to embezzle and if and how they legitimised their actions is hard to say. What was in effect salt stealing can simply be dismissed as a manifestation of the wayward habits of a societal sub-group with strong criminal proclivities. On the other hand, the salters may have felt that they had a customary right to the home market, in which their masters had shown little interest prior to the 1660s. If accounting practices were as sloppy as Tucker suggested, there may have been some difficulty, which was exploited by the salters, in distinguishing salt which should have been delivered to the saltmasters from that which properly belonged to the salters in the form of their 'overplus'. Although factors such as this are likely to have played their part, perhaps a more tangible cause of embezzlement was indebtedness on the part of the salters. Their right to sell their 'overplus' was dependent upon their first delivering to the saltmaster the quantity of salt they had agreed to make in return for their pancoal. For this contracted quantity, the 'taught', they also received modest cash payments, between 3*s*. 4*d*. and 4*s*. Scots the boll for each of the two bolls (and two pecks) required weekly of each of the Wemyss salters in the mid-seventeenth century for example.[94] Unfortunately, only rarely has data survived which includes both a record of the rate salters were to be paid for their 'taught', and how much salt they actually delivered to their masters. Failure to deliver could present serious financial problems for the salters. Wemyss salters who fell behind with their 'taught' salt without reasonable cause, that is, due to storms or pan repairs, lost 30*s*. Scots for each undelivered boll. In addition they also forfeited the small but useful payments they received for the regular delivery of their requisite two bolls and two pecks of salt. 30*s*. Scots was also the sum Wemyss was bound to pay the salters for each surplus boll. As was demonstrated in Chapter One, salt production could be interrupted for a variety of reasons, with the result that the salters' income from this quarter dried up. This was starkly expressed in a financial record from Torry early in January 1678: 'The salters no Allowance...no salt made.'[95]

Although an insufficient number of saltwork account books have survived from the seventeenth century to make it possible to judge the extent of what was termed 'hind' salt, evidence from the early 1700s shows that at some works, salters were clearly heavily in debt to their masters. Early in December 1719 for instance, each of the six master salters at Elphinstone pans was owing 'hind' salt, in quantities ranging from 32 to 813 bolls.[96] In cash terms, James Gilphillan, the salter who was most heavily in debt, owed £540 Scots. The average money equivalent for 'hind' salt was just over £372 Scots, roughly equal to a fairly substantial £32 sterling. Similarly at Winton, in the same period, the accounts reveal that each of the twelve master salters was in arrears, the 'rests' or 'hind' salt totalling 1,224 bolls in the year ending July 1717.[97]

However, what for the greatest part of the seventeenth century was not perceived to be a major problem, became a considerably more important matter as the eighteenth century wore on. For the most part, salt proprietors in the 1600s had turned a blind eye to their salters' depredations of their stock. After all, the greatest part of their own trade lay in volume sales outside Scotland. The salters were concerned only with serving the domestic landsale market which lay within the immediate vicinity of the pans. The seventeenth century too was a world in which there was some blurring of the distinction between customary and property rights, and in which perquisites had not yet come under the attack to which they would be subjected from the more rational capitalist manufacturing ethos of the next century.[98]

Considerable changes in attitude and circumstances began to take place in the later seventeenth and early eighteenth centuries. As was seen in Chapters Two and Four, diminishing markets overseas and fewer opportunities for bulk sales, encouraged saltwork proprietors to look more carefully at how they might best exploit their monopoly of the rather more limited Scottish domestic market. One consequence of this was that the size of the salters' 'overplus' became subject to increasingly strict scrutiny.[99] Later in the century too a number of the industry's leaders evinced greater concern than their predecessors had about the efficiency with which their works were managed and paid more attention to 'oeconomy', profit margins and detailed and verifiable accounting techniques.[100] The introduction of salt officers from 1713, and their concern to account for virtually every grain of salt which was made in the pans and sold from the works, placed a further restriction on the salters' former opportunity, both by their own skill at the pan and abilities as salt vendors, of achieving a degree of self-respectability as quasi-independent producers. A further blow was delivered to their independence in 1735, in the shape of the aforementioned act which made it illegal for anyone concerned with a saltwork 'to pay the Wages of any Servants employed...in Such Salt Works in Salt'.

During the eighteenth century and until 1823, when the duties on Scottish salt were removed, salt embezzlement in the saltmaking communities was endemic. In the Prestonpans customs precinct, for example, there were 106 seizures of salt by officers in the 12 months from September 1715. In their disregard for the law however, the salters were not alone, with grieves and even the occasional saltmaster evidently not being unduly concerned by the prospect of being taken to court and fined for removing and selling salt from the girnels illicitly.

This was especially true of the years immediately following the imposition of the unpopular salt duty in 1713. Between 1719 and 1721, the enterprising William Morison, of Prestongrange estate, was twice found guilty by the justices of the peace for East Lothian of 'running salt', verdicts which cost him some £430 Scots.[101] In 1724, action had to be taken in the Court of Exchequer against the then Earl of Wemyss, who had evidently been shipping salt from Methil without informing the local salt officers.[102] In the main, however, the market in 'run' salt from the pans was in the hands of the salters. Generally the quantities involved were small, with the typical 'seizure' of salt by customs officers being less than a handful of bushels. With salt tax standing initially at 1*s*. per bushel, which roughly doubled the legal selling price of salt, the sums of 1*s*. plus which could be obtained by the salters for each clandestinely sold bushel, would have provided more than useful supplements to their incomes. For salters burdened with debt such sales are likely to have been essential. Sharp increases in duty in the later eighteenth century provided them with even greater incentives to steal salt from the pans and sell it privately, mainly to the 'poorer sort' of people. So high were the salt duties at the end of the century, and so widespread and deeply rooted amongst the salter communities had embezzlement become, that the saltmasters, as a body, petitioned the Excise Board to urge that more officers be appointed. A salter, they claimed, with 'the active collusion of a wife and family', could, 'by the clandestine abstraction of salt', make 'not less than half a guinea a day'.[103] This was more than twice what they could earn from a day's legitimate work at the pans.

While the latter sum may have been inflated by the frustration the saltmasters felt at their inability to break the 'early habits of dishonesty' they saw in their salters, there is ample evidence to suggest that whatever the costs, in fines, imprisonment or the breakdown of what elements of the paternalistic order had survived into the eighteenth century, salters evidently considered that the gains to be made from illicit sales of salt were sufficient to outweigh the risks of being apprehended and punished.

Few salt officers appear to have carried out their duties with the degree of dedication and courage which would have been necessary to stamp out salter embezzlement. Not all officers were honest or fit enough for the job,[104] while for most, life within the saltmaking communities can only have been tolerable if they were prepared to turn a blind eye to many of their neighbours' habits and follow less than diligently the impassioned orders of their superiors in Edinburgh. Only with the greatest difficulty could they, or the saltmasters and their grieves, have broken down the sort of collective loyalty of the salters which was seen at Wemyss in 1773, for example, where all but one salter had evidently disregarded attempts to stop them stealing. Partly for breaking ranks with his fellow salters and his 'Honesty in the delivering of his Salt while the rest did Smuggle', George Stirling was paid a bonus of additional coals.[105] At the same time the rest of the salters were warned that they should not 'Expect or demand' their winter allowances of

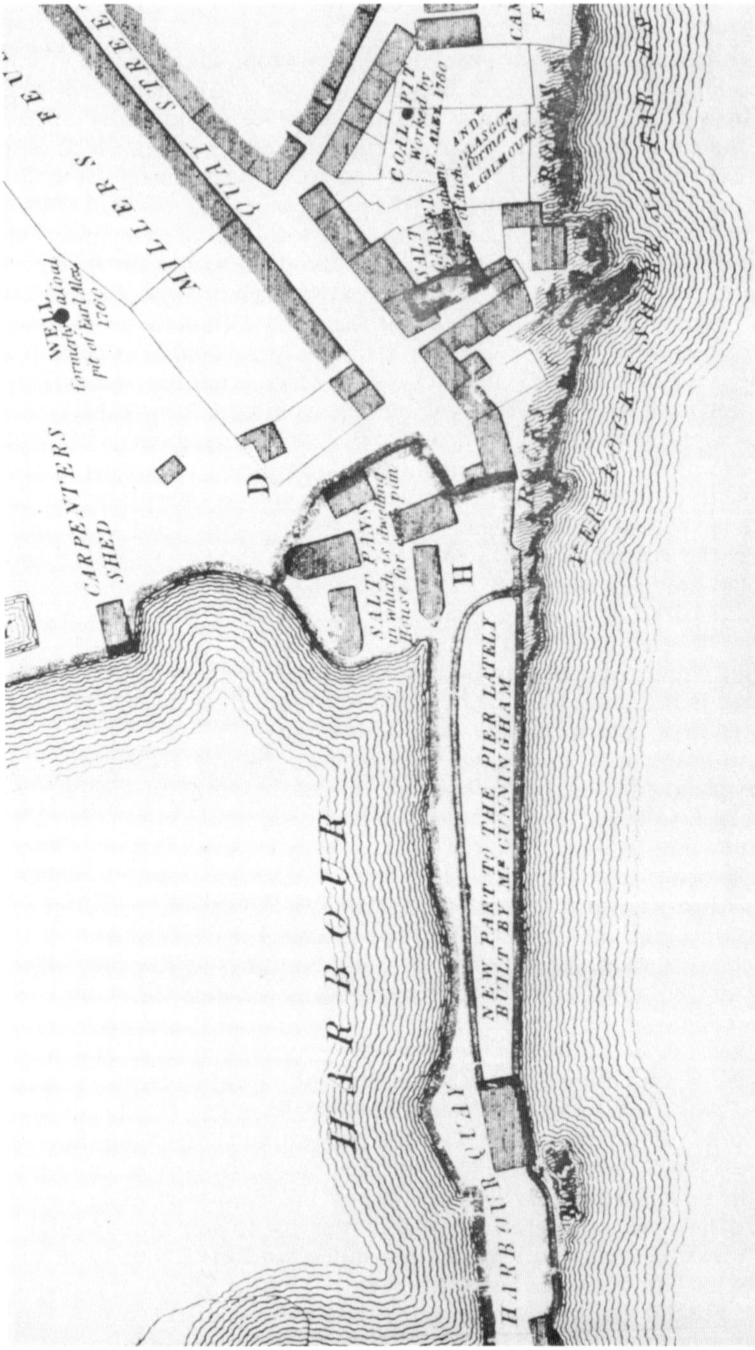

12 Map, late eighteenth century, showing in the centre the saltpans at Saltcoats, Ayrshire. Note that the salters were housed in the upper storeys of the panhouses at these west coast pans. It is worth pointing out that the saltworks (and hence the saltworkers) were not geographically isolated, but located in the heart of this industrialising port town. *Scottish Record Office.*

extra coal, until they swore separately that the salt in their pans would not be 'Abstracted, Smuggled or Run'. Such was the nature of salt manufacturing technology however, and the process by which salt was made, that only attendance at the pans for twenty four hours a day, seven days a week, would have ensured that salt was not removed illicitly. Belching forth smoke and steam for periods of over twenty four hours, the stone-walled and substantially roofed panhouses, almost bereft of windows and natural light, provided superb cover for salters intent on clandestine activity.

To evade the salt officers was judged by the salters to be an achievement worthy of considerable self-satisfaction. According to one nineteenth century local historian, 'they gloried in it as an indispensable qualification, boasting with much self-complacency that they could hoodwink the guager, and smuggle in his very presence'.[106] In this respect they may well have considered themselves a 'race apart' and indeed only intimate acquaintance with the salters' habits and a pre-dawn rise on the Monday morning would have enabled the salt officers to eliminate the manufacture and clandestine sale of 'Sunday salt'. To avoid having to start their pans from cold on Monday mornings, it was the widespread practice of salters, when leaving the pans on Saturday afternoon, to set a slow fire beneath them. At the same time however, to avoid burning the pans, they also added some fresh brine, which, due to the slower than usual pace of evaporation, produced a highly desirable, large-grained salt.[107] The salters removed this salt from the pans either late on Sunday evening or early on Monday morning. How much of it was made is unknown, but what is clear is that most of it was still being sold privately by the salters, without payment of duty, as late as 1808, when the officers were urged to renew their efforts to keep a closer watch on what happened to the salt which was made on Sundays.[108] However, the practice of keeping the pans going on Sundays and the salters' right to the salt produced during the Sabbath, were long-established traditions, possibly going back to the 1640s, when a number of kirk sessions had begun to make a sustained assault on those who continued to make salt on the Lord's Day. Although the 2nd Earl of Wemyss had instructed his salters in 1660 that their pans should go cold before they attended church, the arrangement which had been arrived at earlier and had presumably suited Wemyss when demand for salt had been intense, was the one to which the salters had evidently decided to adhere. This kept them away from the pans during the daylight hours on Sundays, but allowed them to keep the underpan fires lit.[109]

In other ways too the salters attempted to deceive the salt officers and enhance their unseen and unrecorded earnings. Salt was taken from the pans after the 'second water' had boiled down, when the officers would not have anticipated that salt would be removed until two or three more boilings had been completed. In order to conceal what they were doing, the salters, it was alleged, kept the pan 'in a strong boiling state...and the house full of steam'.[110] Another practice was to keep the salt draining in the 'drabs' in the panhouses for longer periods than were necessary and then moving it to the girnels 'under night', when officers and watchmen were off duty and unable to prevent 'abstraction'. Salters too were in the habit of retaining some of the salt bound for the girnels, claiming, fraudulently, that it was 'foul' and unsuitable for use—and then selling it themselves.

Although it is practically impossible to estimate the proportion of salters' total incomes which sales of embezzled salt formed, except perhaps to say that it was large, there is sufficient evidence from saltwork account books and similar sources to allow firmer statements to be made about the level of their legitimate earnings, although it has to be stressed that there is an almost complete lack of data which would allow for the construction of wage series over long periods of time. The evidence is patchy. The picture also tends to be slightly clearer after 1735, when the 'overplus' was more likely to have been recorded and payment for it made in cash (or meal) than was the case beforehand. Prior to making even tentative statements about salters' visible incomes though, several important qualifications have to be borne in mind.

First, the performance of individual salters varied greatly, as was pointed out in Chapter One. The consequent differentials in income which this produced could be further emphasised at works such as Grange in the later 1770s, where the salter producing most salt (above a target of 80 bushels) was paid an additional 2*s*. weekly, the second best performer 1*s*. 6*d*. and the third best, 1*s*.[111] In addition, it has to be emphasised that there were often inexplicably large local and regional variations in payment rates which make the search for typicality something of a lost cause. Secondly, while estimates of average weekly earnings are useful for purposes of comparison with other groups of workers, they grossly distort the actual income patterns of saltworkers, which were highly irregular.

The concept of annual average earnings would have meant little or nothing to the Scottish saltworker. Production was frequently interrupted and earnings from salt output consequently fell or ceased altogether. Nevertheless, at most works, salters received some compensation for their losses of earnings during periods of temporary unemployment, although there were exceptions, as in the case of Lord St Clair at Dysart in the mid-seventeenth century, about whom it was reported that during inclement weather or when the pans were damaged, he gave his salters, 'nothing but his goodwill'.[112] Furthermore, it should be remembered that the work load of the salter is not easily compared with that of other groups of workers. Basing calculations on the 'typical' working week of a salter, when three or four draughts of salt were produced, a nominal total of 72 or 96 hours results. However, salters were not actually hard at work during the whole length of time it took to make a pan of salt, but enjoyed interludes for leisure, rest and even sleep.

Rather than simply handing out sums of money or meal however, proprietors usually preferred to employ the salters on a casual basis on tasks associated in some way with the upkeep of the works. Payment rates for these tasks through the eighteenth century (little data on this for the earlier period has survived) were the same as or marginally below those paid to day or casual labourers. At Limekilns during 1713, for example, Sir Peter

Halkett's six salters were each paid 6*s*. Scots per day for each of the five days they worked at 'redding', or removing rubbish, from the bucket pots.[113] This sum, the sterling equivalent of which was 6*d*., was a fairly standard daily rate of pay for salters engaged on casual work for much of the eighteenth century. Although bucket pot cleaning and repair were the tasks to which salters were most commonly set, they were also often employed in assisting the smiths and masons who were working on their pans and panhouses. At the turn of the nineteenth century the daily rate at Grange for 'labouring with the smiths' had risen to 2*s*. daily. This was roughly half the rate then being paid to working colliers. Payments for idle periods continued to be made later in the nineteenth century, although salters were far less likely to be paid on a daily basis. Their effort was more closely monitored and in common with other industries, task payments were often instituted—1*s*. for each of the four salters that were usually required to clear a bucket pot at Dysart in 1817, for instance—while on other occasions at the same works, hourly rates of 1½*d*. were paid. At Pinkie in May 1847 'Thomas Forrest & Co.', salters, earned 5*s*. on piece work rates for laying pan ashes on the bed of a railway track.[114]

Although most stoppages in salt production were 'involuntary', a small but significant number were clearly 'voluntary'. That is, the workers concerned clearly opted for leisure rather than work-derived income, or considered that obligations beyond the requirement to attend the pans should take precedence. Such absences may also serve to strengthen suspicions about the prominent role in salters' living standards which embezzled salt may have played. Occasionally too, salters were sheltered from total dependence upon legitimate saltwork earnings by having a foothold in the rural economy. John Carnochan, a salter at Turnberry saltpan in Ayrshire in 1754, also described himself as a thatcher. Indeed, further weakening the notion of a sub-class of outcast salter serfs is the evidence from Prestongrange estate in the early eighteenth century, which reveals that John Fleuchart, a salter, was also a feuar of half a tenement and a piece of waste ground, land for which he was bound to pay an annual sum of money, some hens and six days harvest labour. His neighbours and fellow minor landholders comprised a wide cross-section of Scottish lowland society, from weavers, masons and colliers, through proprietors of single saltpans to surgeons and the most substantial feuar on that estate, James Erskine, advocate.[115] And in spite of what has been written about their allegedly brutish characteristics, as Tables 1.2 and 1.3 showed, familial concerns could keep salters from the pans, as they tended close relatives who were sick, mourned the death of mothers, brothers or wives and got married. Salters such as David Greenhill and Andrew Adamson of Dysart, who abandoned their pans during a 'Fast Day' in 1817, were clearly not prepared to neglect their spiritual needs. As has been seen, salters also participated in local fairs and national holidays and celebrations, although rarely did such festivities account for such a high proportion of 'silent' weeks as they did at Winton in 1717–8, when they accounted for 26 per cent of the total weeks lost.[116]

While both regular as well as unpredictable cessations of work produced periodic dips in the pattern of earnings of salters, the payment of premiums and bonuses provided them with brief interludes of considerable prosperity. In the seventeenth and eighteenth centuries it was usual for master salters to be paid annual, half-yearly or quarterly bounties, which were part and parcel of the contractual arrangement between the salter and his saltpan proprietor. In 1654, the Earl of Wemyss was paying his master salters quarterly bounties of £7. 3*s*. 4*d*. Scots, although from this sum they were expected to pay for their 'poles and shovels'. Substantially smaller sums, 29*s*. Scots, were paid annually as 'earnest money' to the master salters at Winton, although as was seen earlier, these payments coincided with and were supplemented by a saltmaster-sponsored bout of heavy drinking. Later, as at Grange, the payment of an annual bonus (of £5 to be shared between five or six master salters at Grange in 1779), could be dependent upon the salters' achievement of a certain output target.[117] Perhaps reflecting the industry's greatly improved fortunes during the French wars at a time when life binding was virtually all but a distant memory amongst salt workers, each of the master salters at Pinkie in the later 1790s was being paid 18*s*. 'setting money' twice yearly. This was roughly equivalent to a coal miner's weekly wage and therefore represented a considerable 'windfall' payment to salters. However, by the 1840s and 1850s, when rapid population increases and the influx of Irish immigrants had greatly swelled and weakened the labour market and the salt industry was but a pale shadow of what it had once been, such payments appear to have been discontinued.

Part of the payments made to master salters, however, had to be repaid to their assistants. Where the working unit was the family this of course did not apply: the master salter's income was presumably the family's income. Where the master salter was employing assistants from outside his immediate family though (and where this occurred is not always clear), enormous difficulties arise. It was rare for bonus payments which were tied to salt output to be made by saltmasters directly to salter assistants: how much they received was usually determined by the master salter. Only exceptionally did the saltwork owner or lessee take a hand in this, as occurred at Methil in 1772, when a winter allowance of additional coal was paid to the master salter, along with an extra 3*s*. which was paid to each of the salter assistants and 2*s*. 6*d*. to the salter boys.[118] As has been said already, the weekly or fortnightly wages of the assistant salters were normally paid by the master salters. Unfortunately, very little hard information is available about what proportion of production bonuses, or indeed of any other payments, master salters made to their assistants. Furthermore, as only a few scraps of evidence concerning payments made to salter assistants have survived, all of which are drawn from the second half of the eighteenth century, it is only possible to engage in a very limited discussion about their wage levels.

What the available data suggests, is that assistant salters were paid at or below the level of day labourers. At Grange in 1780, for instance, the master salters were evidently paying their assistants 1*s*. 6*d*. for every draught of salt taken from the pans. Hence even in good weeks, assistant salters can hardly have been earning more than between 4*s*. 6*d*. and 6*s*. [119] On top of this sum, however, an allowance has to be made for the provision of housing. With their masters, they could probably also engage to some extent in the black market for Scots white salt.

Most visible earnings figures given for master salters then, will be deceptively high. A partial exception may be Ayrshire, where it appears that the responsibility for paying the 'blowermen' fell to the saltmasters. These were men who appear to have partly supplemented the assistance which master salters obtained from members of their families. Their principal function was to maintain the underpan fires. Like the salter assistants, they were poorly paid, at the rate of 2*s*. a week in the 1750s, and 5*s*. weekly in the 1770s.[120] In the long term, the master salters' wage rates were linked to the price of and demand for salt. These factors were generally more favourable from *c*.1570–1670, *c*.1690–1730 and between *c*.1793 and 1823. In the intervening periods it is likely that the salters did less well.

Although no earnings figures are currently available for the salters of Wemyss and Methil during the lengthy periods of prosperity during the seventeenth century, other contractual payments made to them when they were not making salt may be used as a crude measure of the value their masters attached to them. They may also indicate roughly the levels of earnings salters may have been achieving when their pans were going. In 1660, for example, the Earl of Wemyss was prepared to pay his master salters compensation of £3 Scots a week when the pans were being repaired.[121] In addition they were then being paid a quarterly bounty fee of £5 Scots. Assuming regular receipt of sums of this order, their annual earnings figure, excluding any income from private salt sales, have been around £14 sterling, which is certainly comparable with what some Scottish coal miners may have been earning in the early eighteenth century.[122]

That the salters could achieve highly respectable earnings from their work at the pans is also suggested by figures taken from the accounts of the Duchess of Hamilton's saltwork on Arran. David Wannan was a master salter who was brought over to the island in February 1711. He could not begin to make salt immediately though, and was therefore paid at a weekly rate for two weeks. For these he was paid a total of £15 Scots, £7.10*s*. Scots for 'a wicks wages from His own House to Aron', and the same sum, for the second week, for the work he did on preparing the pan.[123] This rate was higher than that paid to anyone else at the joint saltwork-colliery venture, and was, for example, more than double the £3. 12*s*. which was paid weekly, for six days work, to the skilled pit sinker, John Baird. Wannan's treatment does not seem to have been exceptional. In 1712 and 1713, one of his successors was being paid for making salt at £1 Scots per day, while, unusually, his assistant received a separate 10*s*. Scots. Between January and April 1714, the master salter Malcolm McMillan's weekly earnings averaged some £25 Scots, although MacMillan was required to pay the wages of his assistant from this sum. The hewers who were cutting the coal for the pans were then being paid between 10*s*. and 14*s*. Scots per day.[124]

Short-term fluctuations apart however, the price of salt did not stray much beyond an average of between 8*d*. and 1*s*. 1*d*. per bushel, before duty, until the 1790s. As the industry's output too, failed to rise above its early eighteenth century levels, it is not surprising to find that the wages paid to master salters changed little until the later decades of the century. In the 1740s and 1750s, for instance, it is difficult to find examples of salters whose visible earnings were any more than 10*s*. or 11*s*. per week.[125] Earnings such as these are above the lowest figures given for Scottish colliers, but some way below the highest.[126] There may have been some slight advance during the upturn of the late 1760s and early 1770s. Over a 20 week period in 1777, for example, the master salters at Craigie pans earned an average 13*s*. 6*d*. weekly, although they were paid every fortnight.[127] By the end of the century however, in common with other groups of workers, the salters had made substantial gains in their money wages. In the 12 months from November 1791, the average weekly earnings of the four master salters at Pinkie were 18*s*. 6*d*, a sum which was made up of three parts: how much salt they made, at 3*d*. the boll, a bounty, and a 21-part share in the salt sold from the girnels. Their annual earnings ranged from £43, in the case of John Miller, to William Walker's £52. That earnings were of this order, and higher, in the 1790s, is confirmed from other sources.[128] Although for reasons which have partly been explained, comparison with other workers is a hazardous undertaking, it does seem worth noting that the salters' earnings appear to have been comparable with the upper range of those obtained by Scottish colliers, with whom they shared the requirement to pay their assistants, whether as members of their family, or as paid labourers.[129] Even so, their earnings were clearly above the average for agricultural labourers and roughly on a par with cotton spinners.[130]

Although the work pattern of the salters was more regular than it had been earlier in the century, there were still huge fluctuations in their earnings. In the week ending 12 May 1792, for instance, James Christie, at Pinkie, made 11*d*. On other occasions he was able to make just under £2. Salters, 'when they work', noted John Girvin, could, in 1796, make 20 bushels of salt in a day, which, at 3*d*. per bushel, could produce a highly respectable daily wage of 5*s*.[131] Technically, of course, it was impossible for salt manufacturing to go on unchecked. The salters, it appears, also valued some time away from the pans. They were, nevertheless, encouraged with financial inducements to work hard. At Grange in 1801, each of the master salters had to turn out at least 15 bushels of salt, during the summer, in return for 10 carts of coal. Thereafter they were paid 9*d*. per bushel, with a 3*s*. bonus should they make 70 bushels or over.[132] If they made this perfectly feasible quantity in a week, their earnings would have been at least £2. 4*s*. 3*d*.

With the end of the French Wars in 1815, however, came the beginning of a long period of relative decline for the salters. In 1816, the costs of saltmaking at Dysart were being calculated on the basis that the salters received just over 3*d*. for each bushel they made. This was the price they had obtained in the mid-1790s.[133] Between May 1817 and February 1818, no master salter at Dysart was able to make over £2 from his pan, and indeed the average weekly earnings of the six master salters were in the region of £1. 5*s*.[134] Some, such as Thomas Hay, at the Wicket Pan, made much less than this. By mid-century, the salters' earnings at Pinkie had dropped back to the levels of the early 1790s. From the beginning of August to the end of December 1845, for instance, the average weekly earnings figure was 18*s*. 6*d*., paid fortnightly.[135] With the removal of all taxes on salt in 1825, the incentive for consumers to buy salt illegally disappeared. For the salters, the ending of this trade was a grave loss, sending them even more certainly down the road to poverty and lowly social status. Their descent may have begun by July 1823, when a riot, in which several people were hurt and one person was killed, broke out on Mussselburgh Links. The outcome of a dispute over gambling, prominent parts in the violence were allegedly played by Irishmen, colliers and salters.[136] With the decline of the industry after 1825, in much smaller numbers, the Scottish salters joined the handloom weavers as victims of the competitive forces which were unleashed during the Industrial Revolution. In their case however, anything which had approached a 'Golden Age' had been tarnished by the backcloth of serfdom, and in sharp contrast to the handloom weavers, posterity has paid little attention to the passing of their trade. The salters' loss, however, was the Scottish economy's (marginal) gain, as workers released from the industry turned to 'more profitable work for which Scotland was better-endowed.'[137]

6

POSTSCRIPT: SALTERS AND THE 'EMANCIPATION ACTS'

Much has been written about the reasons why a system of serfdom was instituted amongst the Scottish collier and salter populations in 1606, and how their legal servitude was brought to an end in the late eighteenth century.[1] In the main, however, the arguments developed have been based on what has been understood about the requirements of the coal mining industry and of the coal miners' circumstances: the implication has been that what applied to the miners equally applied to the saltworkers. Closer examination, however, reveals that such an approach has significant weaknesses, and indeed, on some matters, seems simply to have been misplaced as far as the salt industry and the saltworkers are concerned.

Three factors are generally considered when explanations are being sought for the creation of a system of legal servitude in the coal industry. Most important was 'Nef's Revolution', when, during the seventeenth century, coal mining experienced dramatic growth which necessitated additional labour supplies. Labour, however, was difficult to recruit and hold. It is also thought that the Scottish Poor Laws of 1579 and 1597, which allowed vagrants to bind themselves for life to any willing employer, may have been used by coalmasters as a means of securing hewing labour, and can be seen as forerunners of the legislation of 1606.[2] That coal (and salt) were 'services necessary for this kingdom', to use Viscount Stair's phrase of 1693, may also have played a part in influencing the legislators.[3] The result of these forces was the Act of 1606, the principal elements of which were outlined at the beginning of Chapter Five. As the salt industry too was experiencing a period of intense demand at the turn of the seventeenth century, it seems reasonable to suppose that saltmasters would also have been anxious to limit the movement of their skilled salters in the manner allowed by the 1606 Act.

It seems highly unlikely, however, that the saltmasters would have been encouraged to seek such a measure on the basis of the precedent provided by the Poor Laws. As has been stressed earlier, it was skilled and experienced saltworkers the saltmasters sought, not beggars, orphans and other unfortunates. It seems far more likely that saltmasters would have been concerned to hang onto the salters and their families they already employed. By restricting the rights of new masters to entice salters away from their current masters, who may have invested considerable effort in training them, it was probably hoped to exert some control over labour costs, which are likely to have been increased by competitive bidding.

It is, however, where the so-called 'Emancipation Acts', of 1775 and 1799, are concerned that the habit of considering the colliers and salters as inseparable must be reviewed more critically. Although humanitarian impulses are held by some to have played a part,[4] the principal motivation of those who wanted to alter the laws and abolish the Court of Session-backed estate customs which governed the conduct of labour in the Scottish coal and salt industries, was a desire to expand the labour supply, a shortage of which was acting as a severe constraint upon their growth.[5] The 'stigma of serfdom', it has been argued, deterred people from entering these trades and thus the Act of 1775 was fought for and passed to begin the process of dismantling the system of life-binding. This was to be done, first, by releasing, in stages, according to their ages, colliers and salters who were already bound, and secondly, by abolishing life-binding for newcomers to the two industries. The 1799 Act has been seen as the second and final part of that process, although recently, greater stress has been placed on its anti-combination clauses.[6] The effect of the 1775 Act on the salt industry, it has been suggested, was to make possible an almost immediate increase in salt production, presumably through an expansion of the size of the labour force.[7] It was not until the 1830s that the bottleneck in the supply of labour for the coal industry was overcome.

Attractive as the hypothesis of a labour shortage may be in the case of coal mining, it holds less water when applied to the salt industry. There is, for instance, little evidence to support the idea that the industry's growth was seriously constrained at any time during the eighteenth century by an absolute shortage of labour. It is true that saltmasters could occasionally be frustrated by the absence of assistant salters, but while they tended to leave the pans with far greater frequency than did master salters, they could usually be replaced relatively easily. As was argued in Chapter Two, it was above all changing market patterns and the relatively slow growth of demand in Scotland which were the major economic determinants of the industry's fortunes in the period. Earlier in the century, there had been occasional local short-term shortages of labour, in the 1720s for instance, but these, it seems, could largely be met from within the salt industry. Indeed, with the slight downward trend in sales which occurred during the period from *c.*1730 to *c.*1760, all that was required for the industry to be amply provided with skilled labour was for currently employed saltworkers to remain within it and ensure that their children replaced them.

It is striking that no cases of disputed ownership of salters have yet emerged from eighteenth century Court of Session records. This contrasts sharply with the coal mining sector, the leaders of which made frequent forays into the legal arena in order to establish their rights over their colliers. Yet instances of master salters leaving their pans without permission are exceedingly rare. It might be thought that it was more difficult for them to move to other saltworks without being discovered. Certainly there were less opportunities for salters to move within the industry than was the case for colliers. There were, after all, only as many master salters' jobs as there were pans in Scotland; 164 at the turn of the eighteenth century, 118 in 1787.[8] Even allowing for the fact that occasionally there were more than two salters per pan, the total number of workers directly involved in salt production full-time (excluding grieves, and labourers and those workers who were employed on the many other, mainly temporary, jobs around the pans) can hardly have been more than 350 at any time during the 1700s. In his *Analysis of the Statistical Account of Scotland,* published in 1826, Sir John Sinclair reckoned that there were only 205 saltworkers north of the border.[9]

Whatever the precise figure, it does seem clear that saltworkers would have been much more prominent than coal hewers, the master salters especially so. The whereabouts of saltworks were well-known, and in spite of the difficulty grieves may have had in monitoring day-to-day operations in the panhouses, it would have been much easier for a saltmaster to hunt down fugitive salters than it would have been for a coalmaster to scour the countryside looking for absconders at every small colliery. The numbers of colliery operatives hugely outnumbered those employed in salt. At Bo'ness colliery and saltworks, for instance, in February 1771 there were 278 people employed in and around the five coal pits, 104 of whom were colliers and miners. At what was, with Wemyss, the fourth largest saltworks in the country, there only 28 salters and their servants. On Preston Island in 1813 , there were at least 50 workers employed in and around the two coal pits: a total of five workers was considered sufficient to operate the two saltpans.[10]

On the other hand, it can be argued that master salters tended to remain at the pan or pans at which they had initially entered for more positive reasons. As has been suggested, while much of the work was physically demanding and carried on in wretched conditions, both the social standing and earnings achieved by master salters may have been considerably higher than has previously been assumed. They were, after all, small masters of men and masters of a trade. And while the conditions of the assistant salters were evidently poor and liable to cause their disappearance from the pans, at least some of them could one day expect to become master salters. Where a master salter's assistant was his son or other close relative, this was clearly even more likely. It was coal mining in the eighteenth century that was an unpopular occupation, throughout Britain, not salt manufacturing.[11] Indeed there is a suggestion that while colliers may sometimes have become salters, there was little movement in the other direction.[12]

What then of 1775 itself and the pressures which built up from the early 1760s for alterations in the labour laws? Whatever it was that kept the agitation going during the critical months of 1774 and early 1775, it was certainly not a shortage of salters. In a letter he wrote in March 1774, Walter Scott, the legal agent of the reformers, told the Lord Advocate that they were 'a species of Servants much easier procured than Coalliers'.[13] The post-1772 recession in the Scottish economy threw salters as well as colliers out of work.[14] Indeed, by as early as February 1771, the number of workers at Bo'ness saltworks had been slashed from 75 to 42. The number of master salters and their assistants was cut by 13.[15] Reinforcing the argument that the industry's leaders are unlikely to have been concerned to increase the labour supply was the formation in 1773 of the Salt Association, which was established not to raise salt output but to reduce it. Of the salt producing precincts of the Forth, it was only Prestonpans which increased its output, largely owing to its unique dependence upon the local domestic market, demand from which increased with the rising population of Edinburgh and the Lothians.[16] If there was any concern about the salters at all, it lay in a somewhat different direction: in the words of one of the reformers, Patrick Robertson, of Glasgow, changes in the law were necessary for 'regulating the conduct of so refractory a race of men as Coaliers and Salters generally are'.[17] Labour control was the issue, not labour supply.

Robertson, however, was speaking on behalf of the Glasgow coalowners. Neither he nor the other west coast coalowners who were vociferously agitating for the end of life-binding had much interest either in the salt industry or the salters. Indeed, from beginning to end, the campaign was dominated by men whose interests lay primarily in coal.[18] Furthermore, it seems that they were less interested in numbers as such, and far more concerned about finding means by which they could exercise greater control over their 'refractory' colliers and reduce what they considered to be their unreasonably high wages. Order would be restored, they hoped, and colliers' wages cut, if the effectiveness of collier combinations could be reduced. Hence freedom from life servitude was conditional upon those who wished to go being able to satisfy local magistrates that they had not recently been involved in combinations.[19] The outlook for coal, however, was perceived to be good, provided that coalmasters were able to increase coal output and compete in price terms, mainly by reducing their labour costs. The saltmasters, having long abandoned any hopes of exporting salt, were not subject to the same frustrations as the coalmasters and accordingly, played little or no part in the repeal agitation.

While many Scottish colliers were still bound for life by 1775,[20] the extent to which salters were similarly constrained in practice, is, in the absence of legal records, by no means clear. It is tempting to suppose that the system of life binding for most salters may simply have withered away. In the accounts of Bo'ness colliery for 1770 and 1771 for instance, a distinction was made between 'free' and 'bound' colliers. No such classifications were used for the salters and there is no suggestion that they were bound for life.[21] The widespread payment of annual sums of 'earnest money' made to master salters and sometimes to their assistants from at least the middle of the seventeenth century, may indicate that on their own, estate customs concerning life binding and the Court of Session's interpretations of the rights of coalowners and lessees, which became increasingly onerous for colliers in the eighteenth century, may have had to be underpinned by additional means where the salters were concerned. After all, unsuccesful attempts were made to impose serfdom on other groups of workers.[22] In theory, serfdom was an iniquitous system: the statute book, however, is not the best source for ascertaining how it worked in practice.

It is certainly quite erroneous to say that the Act of 1799 completed the process of emancipation for either the colliers or the salters. The Act, 'to explain and amend the Laws relative to Colliers in that part of Great Britain called Scotland',[23] makes no mention of either the salt industry or its workers. Passed following mounting concern amongst east coast coalmasters about the spread of combinations and rising wage costs,[24] the Act had nothing to say about life servitude. Instead it prescribed means by which wages might be better regulated and extended the general anti-combination laws to include colliers.

If there is room for reasonable doubt about the extent to which the burden of servitude was shared equally by the salters and colliers, paradoxically and in different ways, what they were both to suffer under from the 1820s and 1830s respectively, was the subjugating power of the free market, allied in the latter case with the markedly more vigorous attitude to labour relations which was adopted by the ironmasters, who in the ninteteenth century came to control great sections of the Scottish coalfields.[25]

APPENDIX
A NOTE ON WEIGHTS AND MEASURES

Lack of uniformity of weights and measures plagues the historian anxious to measure accurately output, trade or consumption patterns in Scotland from earliest times. Significant regional and local variations were still in evidence in the mid-nineteenth century. Valiant attempts have recently been made to make sense of the variety of measures, notably by R E Zupko and T C Smout and I Levitt (*Scottish Historical Review,* LVI, October 1977). This brief note attempts to add briefly to those contributions and to impose some order upon the apparently anarchical range of measures used by salt producers.

From the later sixteenth century the Scottish saltmasters (at least those on the east coast) do appear to have adopted, nominally at least, three regionally distinct sets of salt measures. These were apparently confirmed, in order to provide merchants with some degree of certainty about the quantities of salt they were purchasing, in 1642 or 1643 and should have been used from January of the latter year. Saltworks at and around Culross and 'all above the Queensferry' were supposedly bound by the Culross measure. The Culross salt chalder comprised 21 bolls of four firlots of 29 pints each. In the vicinity of Prestonpans the chalder comprised 18 bolls of four firlots of 20 pints. Thirdly, there was the Wemyss measure, which was the nominal standard for the north side of the Forth from Queensferry to Fife Ness. The Wemyss chalder evidently comprised 17 bolls of 74 pints. The firlot was 26 pints. In the early eighteenth century the Fife salt chalder was considered to weigh 60 bushels, or one and a half tons.

In the sixteenth and seventeenth centuries the pint appears to have been the common unit of measurement. Unfortunately however, for a variety of reasons, some of which have been outlined by Zupko, the pint measure was by no means uniform. Indeed at Wemyss the pint used when salt was being delivered to merchants was bigger than the pint measure which was used to determine how much salt the salters should return to Wemyss for their coal. Only 22 of the salters' pints comprised a firlot. Another complication arises with the boll, which appears to have been comprised anything between three and over four bushels. The measures were not wholly arbitrary though: there was apparently a clear distinction between the 'small' and the 'large' boll. In saltwork account books of the period however, it is not always clear which boll measure was being used. From Pitfirrane, in Fife, comes evidence of the size of the respective bolls in that location (and perhaps for Fife) as they stood in 1716. The small boll weighed three bushels nine and a half pounds; the Imperial equivalent of the large boll was reckoned to be four bushels ten pounds. On the other hand, when allowance was made for wastage in the girnels, the large boll (a salt water measure) was judged to weigh only 226 pounds, whereas four bushels weighed 224 pounds.

In spite of the very real difficulties which would face an aspiring quantifier should more evidence concerning salt output and sales pre-1713 come to light, or if that which is available was thought to be capable of revealing in more detail (or perhaps of challenging) the pattern of salt sales outlined in Chapter Two, some assistance can be probably be obtained from the above. The Culross chalder of 2,436 pints was almost twice the size of the Wemyss chalder (1,258 pints) and considerably greater than the Prestonpans chalder (1,440 pints). Of course while it would be comforting to assume that the regional variations were strictly adhered to, a major problem would be to establish which salt measure a particular clerk or grieve had been using. However, as was said in the Preface, after 1713 the bushel of 56 pounds became the standard salt measure used by the salt industry. Thereafter questions have to be raised about the accuracy of the measurer rather than the measure.

NOTES

Place of publication is London unless otherwise stated

INTRODUCTION

1 D Defoe, *The History of the union of Great Britain* (Edinburgh 1709), 140–1. The extent to which salt and the salt industry were issues at the time of the Union can be also be seen in Defoe's references to salt in his correspondence and by his claim to have temporarily adopted the guise of a partner in a Scottish saltwork. During his stay in Scotland, working on behalf of Lord Harley, he adopted many such roles in his efforts to persuade Scots from a variety of important interest groups of the advantages which union with England would bring them. See G H Healey (ed) *The Letters of Daniel Defoe* (Oxford 1955), 145–6.

2 See F Braudel, *The Structure of Everyday Life: Civilisation and Capitalism, 15th–18th Century* (1981), 198–210; R Tannahill, *Food In History* (1973), 214; 'Salt and the Salt Trade', *Scottish Monthly Magazine* (1836), 1, 129–39.

3 R K Multhauf, *Neptune's Gift: A History of Common Salt* (Baltimore and London 1978), 4; for a stimulating discussion about salt appetite, see D Denton, 'Instinct, Appetites and Medicine', *Australia and New Zealand Journal of Medicine*, 2 (1972), 203–212.

4 See, for example, M Plant, *The Domestic Life of Scotland in the Eighteenth century* (Edinburgh 1952), where salt is hardly mentioned, yet great stress is laid on the place of meal (which was invariably seasoned with salt) in the Scottish diet.

5 T C Smout, A *History of the Scottish People 1560–1830* (1972 ed), 284; N Murray, *The Scottish Handloom Weavers 1790–1850: A Social History* (Edinburgh 1978), 98–102.

6 J Boswell, *Journal of a Tour to the Hebrides with Samuel Johnson* (1936 ed), 331–2.

7 J Girvin, *An Investigation of the Substitutes for the Present Duties on Salt recommended by Lord Dundonald* (Edinburgh 1801), 25; I MacLeod (ed), *Samuel Robinson: Reminiscences of Wigtonshire* (Glasgow 1984), 1.

8 J Walker (ed), *The Institutions of the Law of Scotland* (Edinburgh and Glasgow 1981 ed), 1013.

9 Multhauf, *Neptune's Gift*, 3; A and N L Clow, *The Chemical Revolution* (1952), 46–64.

10 B Lenman, *An Economic History of Modern Scotland* (1977), 124–5.

11 T C Smout, *Scottish Trade on the Eve of the Union 1660–1707* (Edinburgh and London 1963), 237.

12 Scottish Record Office (SRO), Hamilton MSS, GD 406/1/4975, 'Memorandum concerning Coall and Salt', 1702.

13 A Hamilton, *Prestonpans: Scotland's Last Saltworks* (Edinburgh nd), 1; J M Webster, *Dunfermline Abbey* (Dunfermline 1948), 63–4; O Kelly, *The Saltpans at Saltcoats* (Saltcoats 1972), 1–2; C Innes, *Sketches of Early Scotch History and Social Progress* (Edinburgh 1861), 145.

14 S G E Lythe and J Butt, *An Economic History of Scotland 1100–1939* (Glasgow and London 1795), 41.
15 *Register of the Privy Council of Scotland* (*RPCS*), 1st ser, II, 265, 296.
16 *Records of the Burgh of Prestwick* (Glasgow 1834), 29.
17 I A Adams, 'The Salt Industry of the Forth Basin', *Scottish Geographical Magazine*, 81 (1965), 157.
18 F J Shaw, *The Northern and Western Islands of Scotland: Their Economy and Society in the Seventeenth Century* (Edinburgh 1980), 161; J Brand, *A Brief description of Orkney* (Edinburgh 1701), 36–7.
19 For drawing my attention to this site (HY 412065), I am grateful to Graham Douglas.
20 For a detailed study of the Arran site, see C A Whatley, 'An Early Eighteenth Century Scottish Saltwork: Arran, *c*.1710–1735', *Industrial Archaeological Review*, 6 (1982), 89–101.
21 See C A Whatley, 'Scottish salt making in the 18th century: a regional survey', *Scottish Industrial History*, 5 (1982), 16; *'That Important and Necessary Article': The Salt Industry and Its Trade in Fife and Tayside 1570–1850* (Dundee 1984), 38–40.
22 SRO, Hamilton MSS, GD 406/1/4342, Daniel Hamilton to Earl of Arran, 23 May 1698.
23 D Beveridge, *Culross and Tulliallan* (2 vols, Edinburgh and London 1885), 1, 202; *Register of the Dioscesan Synod of Dunblane 1662–1688* (Edinburgh and London 1877), 118.
24 B F Duckham, *A History of the Scottish Coal Industry I 1700–1815* (Newton Abbot 1970), 15.
25 Whatley, *'That Important and Necessary Article*', 24.
26 See M Moss and J R Hume, *The Making of Scotch Whisky* (Edinburgh 1981).
27 J Ellis, 'The Decline and Fall of the Tyneside Salt Industry, 1660–1790: A Re-examination', *Economic History Review*, XXIII, 1 (1980), 53.
28 A L Donovan, *Philosophical Chemistry in the Scottish Enlightenment: The Doctrines and Discoveries of William Cullen and Joseph Black* (Edinburgh 1975), 83; A R B Haldane, *The Great Fishmonger of the Tay: John Richardson of Perth & Pitfour* (Dundee 1981), 18–9; H D Smith, *Shetland Life and Trade 1550–1914* (Edinburgh 1984), 76; C Singer (*et al*, eds), *A History of Technology* (6 vols, Oxford 1958), IV, 44–8.
29 W Ramsay, 'On Culinary Salt; with the Means of Purifying it from Substances which contaminate its Quality', *Transactions of the Highland Society of Scotland*, IV (1816), 378.
30 W Brownrigg, *The Art of Making Common Salt* (1748), 154.
31 Disbursements for blood and eggs appear in many saltwork account books and descriptions of the salt manufacturing process. See for example the 'Miscellaneous Acounts' presented in the legal dispute between S Mitchell and W Orr in 1764, SRO, CS 237/1P/M/3/1; and T J Salmon, *Borrowstouness and District* (Edinburgh and London 1913), 161.
32 Ramsay, 'On Culinary Salt', 379–80.
33 D Crawford (ed), *Journals of Sir John Lauder of Fountainhall 1665–1676* (Edinburgh 1900), 92.
34 A Cochrane (Earl of Dundonald), *The Present State of the Manufacture of Salt Explained* (1785), 5.
35 Girvin, *Investigation*, 16.
36 Cochrane, *Present State*, 9.

37 National Library of Scotland (NLS), Wemyss MSS, MS 5723, Tutors Sederunt Book, 1719–1733; SRO, Dundonald MSS, GD 233/183/Folder 15, 'Report by Messrs Henderson, Grieve etc, as to Forest of Culross, Tullyallan Coal etc', 16 October 1795.
38 R A Houston, 'Marriage Formation and Domestic Industry: Occupational Endogamy in Kilmarnock, Ayrshire, 1697–1764', *Journal of Family History*, 8 (1983), 215–229.
39 NLS, MS 5723, Wemyss Tutors Sederunt Book, 25 April 1719.

CHAPTER 1

1 T C Smout (ed), 'Journal of Henry Kalmeter's Travels in Scotland, 1719–20', in *Scottish Industrial History: A Miscellany* (Edinburgh 1978), 41.
2 H M Cadell, *The Rocks of West Lothian* (Edinburgh and London 1925), 322; see too, Salmon, *Borrowstoness and District*, 161.
3 Duckham, *Scottish Coal*, 26, talks of 'common estimates that some 16 tons of coal were often needed to distil one ton of salt'. I have come across no contemporary estimates of this order; for an authoritative contemporary account of coal consumption in marine salt making, see *Scots Magazine*, 66 (1804), 763; for a modern view see Ellis, 'Tyneside Salt Industry', 46–7.
4 I H Adams, *The Making of Urban Scotland* (1978), 53.
5 Whatley, 'Scottish salt making', 4.
6 Dr Singer, *General View of the Agriculture in the County of Dumfries* (Edinburgh 1812), 427.
7 Whatley, 'Scottish salt making', 6–7.
8 A Fenton, *The Northern Isles: Orkney and Shetland* (Edinburgh 1978), 212–4.
9 *Ibid*, 213; G Low, *A Tour Through the Islands of Orkney and Schetland 1774* (Inverness 1978 ed), 12, 18; for information on salt making sites in Orkney I am also indebted to Graham Douglas for letting me see his field sketches.
10 SRO, Agnew of Lochnaw MSS, GD 154/461/2, Tack, Sir Andrew Agnew and Patrick McNeilly, 1741.
11 SRO, Agnew of Lochnaw MSS, GD 154/461/6, Tack, Sir Stair Agnew and Alexander Martin, 1776.
12 Dumfries Museum, Dumfries Customs Letter Books, 26 April 1714; I am grateful to Dr Ian Donnachie for lending me his transcripts from this source.
13 Singer, *General View*, 529.
14 *Ibid.*
15 *Ibid*, 528; see too 'Account of the Curiosities At Dumfreis by Doctour Archbald', *MacFarlane's Geographical Collections* (Edinburgh 1908), III, 185.
16 Although there were some local variations in the size of east coast pans, see, for example, *Whatley, 'That Important and Necessary Article'*, 22.
17 *Scots Magazine*, 66 (1804), 760.
18 See Note 12 above.
19 Calculated from SRO, E 536/1–3, Salt Charge Vouchers. 1714–17.
20 See Note 10 above.
21 This is treated more fully in Chapter Three.
22 Cawdor Castle MSS, Charter Room, Bundle 656, 'Obligement: James Scott to the Laird of Cawdor, 1694'.
23 *Ibid.*
24 Whatley, 'Scottish salt making', 10.
25 Shaw, *Northern and Western Islands*, 161.

26 For allowing me to make use of his drawings and photographs of this site I am indebted to Raymond Lamb.
27 NLS, MS 3088, Bogie Salt Book, 1720–32.
28 *Ibid*; NLS, Adv MS 80.1.3, Dundas of Dundas, Archibald Robertson to George Dundas, 20 January 1727; for a description of the formal organisational structure of both the colliery and saltwork on the Winton estate at Cockenzie, see SRO, Forfeited Estates Papers, FEP 1715/E 661, Box 196, 'Memoriall To Sir Henry Hoghton one of the honble The Commissioners of Inquire anent the Coal and Salt Books of Winton An:1716'.
29 Wemyss Castle MSS, Oncost Book, Dysart Salt, 1817–8.
30 See, for example, the account of operations at Grange pans sometime prior to 1889. In order to save fuel and time the salter would undertake underpan repairs while the fires were still going, by lying a board on the coals, Cadell, *Rocks of West Lothian*, 322. Salt manufacturing, Cadell reported, went on day and night, six days a week.
31 See Beveridge, *Culross and Tulliallan*, I, 194, 202, 221, 241–2.
32 Calculated from SRO, Forfeited Estates Papers, Winton, FEP 1715/E 661/Box 196, Salt Book of Cockenzie, July-July 1716–7; for a detailed study of these works, see C A Whatley, 'A Saltwork and the Community: The Case of Winton, 1716–1719', *Transactions of the East Lothian Antiquarian and Field Naturalists' Society*, 18 (1984), 45–59.
33 Wemyss Castle MSS, Diary of Earl David, 9 May 1659.
34 Quoted in Whatley, *'That Important and Necessary Article'*, 19.
35 NLS, MS 3720, Journal of the Management of the Coall and Salt Works of Prestongrange beginning from the 2nd of April 1748; Wemyss Castle MSS, Diary of Earl David, March 1659.
36 NLS, Cadell MS, Acc 5381, Bound Volume 2, Memorandum Book, Grange Colliery, 17 January 1781.
37 SRO, FEP 1715/E 661/Box 196, Oncost Book, Winton (Cockenzie) Saltwork, August 1716–July 1717.
38 SRO, Hope of Craighall MSS, GD 377/130/7, 'Report of Robert Bell, Architect, on the Buildings at Pinkie Pans, 30 October 1852'.
39 I A Bowman, 'Culross Colliery: a Sixteenth-Century Mine', *Industrial Archaeology*, 7 (1970), 368; A Cochrane, *Description of the Estate and Abbey of Culross* (Edinburgh 1793), 10.
40 SRO, Rosslyn MSS, GD 164/5/37, 'Account, Genl Sinclair To Alexr Mawer for the Saltpans after the Great Storme 1752'.
41 SRO, FEP 1715/E 661/Box 196, Salt Book of Cockenzie, 1716–7.
42 SRO, FEP 1715/E 661/Box 196, Oncost Book, 1716–7.
43 SRO, FEP 1715/E 661/Box 196, Salt Book, 1716–7.
44 Wemyss Castle MSS, Oncost Book, Dysart Salt, 1817–8.
45 Whatley, 'An Early Eighteenth Century Scottish Saltwork', 98.
46 Tucker, 'Report', 5.
47 NLS, Wemyss MSS, MS 5723, Tutors Sederunt Book, 1719–33.
48 NLS, Cadell MSS, Acc 5381, Box 12/1, 'Agreement with Salters', 1 June 1801.
49 SRO, Henderson of Fordel MSS, GD 172/1107/34, Salt Agent's copy letter book, 15 March 1845.
50 NLS, Wemyss MSS, MS 5723, Tutors Sederunt Book, 1719–33.
51 *Ibid.*
52 *Ibid.*
53 SRO, Biel MSS, GD 6/1254, 'Ane Short Accompt of ye present condition of ye Coall and Salt works at Thorntoune', 1681; and see Chapter Two.

54 Cochrane, *Description of the Estate and Abbey of Culross*, 15.
55 SRO, Dundonald MSS, GD 233/183/170, Folder 15, 'Report by Messrs Henderson, Grieve etc, as to Forest of Culross etc', 16 October 1795.
56 *Ibid.*
57 See A J Robertson, *The Origins of the Scottish Railway System 1722–1844* (Edinburgh 1983), 6, 8; Duckham, *Scottish Coal*, 209–10, 212; C A Whatley, 'The Finest Place for a Lasting Colliery: Coal Mining Enterprise in Ayrshire, *c.*1600–1840', *Ayrshire Collections*, 14 (1983), 70.
58 SRO, FEP 1715/E 661/Box 196, Salt Book, 1716–7.
59 NLS, Fletcher of Saltoun MSS, MS 17679, 'Queries anent the Coals at Campbeltown', *c.*1749.
60 D E C Eversley, quoted in R W Malcolmson, *Life and Labour in England 1700–1780* (1981), 75; for references to illness and disease in Scotland, see, for example, M W Flinn (ed *et al*), *Scottish Population History from the 17th century to the 1930s* (Cambridge 1977).
61 B Didsbury, 'Cheshire saltworkers', in R Samuel (ed), *Miners, Quarrymen and Saltworkers* (1977), 156–7.
62 NLS, Wemyss MSS, MS 7523, Tutors Sederunt Book, 1719–33.
63 Wemyss Castle MSS, Oncost Book, Dysart Salt, 1817–8.
64 B Faujas De Saint Fond, *A Journey Through England and Scotland to the Hebrides in 1784* (2 vols, Glasgow 1907), I, 176; see too Didsbury, 'Cheshire saltworkers', 142.
65 Wemyss Castle MSS, Diary of Earl David, 2 August 1658.
66 Smout, 'Henry Kalmeter's Travels', 41.
67 Whatley, '*That Important and Necessary Article*', 19; H Farnie, *The Fife Coast from Queensberry to Fife Ness* (Cupar 1860), 58, 116; Stephen, *The Story of Inverkeithing*, 97.
68 SRO, Callendar of Crichton MSS, GD 247/Box 84/2/11, 'Valuation of wrought & cast iron work of Salt pans of Alex Paterson, leased from Sir John Callendar', 16 February 1813.
69 *Old Statistical Account*, X (1978 ed), St Monance, 739.
70 SRO, Dundonald MSS, GD 233/183/170, 'Report by Messrs Henderson, Grieve etc, as to Forest of Culross', 16 October 1795.
71 See SRO, Ailsa MSS, GD 25/9/Box 46, which includes a full and interesting inventory of the 'Salt Pann of Turnberrie', 15 May 1733.
72 Cadell, *Rocks of West Lothian*, 320.
73 *New Statistical Account of Scotland*, IX, 187.
74 Wemyss Castle MSS, Diary of Earl David, 3 December 1660.
75 *Scots Magazine*, 66 (1804), 762.
76 See Donovan, *Philosophical Chemistry*, 83–6, 117; Brownrigg, *Art of Making Common Salt*, 8.
77 *Ibid*, 84; for a recent survey of Black's work, see D Daiches, P Jones and J Jones (eds), *A Hotbed of Genius: The Scottish Enlightenment 1730–1790* (Edinburgh 1986), 93–114.
78 SRO, Fairholme Papers, GD 237/143/2, 'Memorial—Dr John Roebuck to his Creditors', 1771.
79 Edinburgh University Library, Gen 874/IU/29, Archibald Cochrane to Professor J Black, 17 February 1786.
80 J Butt, *The Industrial Archaeology of Scotland* (Newton Abbot 1967), 135.
81 For several references to Martin's venture, see Kirkcaldy Customs and Excise Office, Letter Book, Collector to the Board, between 1758 and 1766: Whatley, '*That Important and Necessary Article*', 22–3, 37.

82 Smout, 'Henry Kalmeter's Travels', 44.
83 Wemyss Castle MSS, Diary of Earl David, 2 August 1658.
84 SRO, CS 96/3085, P Warner v R R Cunninghame, 1799, 'Supplementary Report by John Clerk of Eldin and Dr John Roebuck upon the Coalworks of Ardeer and Saltworks at Saltcoats'; see too *Old Statistical Account*, XVII (1796), Prestonpans, 65–6.
85 Ramsay, 'On Culinary Salt', 378; east coast salt may also have cost less to produce, see *Old Statistical Account*, VII (1793), Stevenston, 19.
86 See, for example, J MacKinnon, *The Social and Industrial History of Scotland: from the union to the present time* (1921), 18; although referring specifically only to the colliers a similar perspective is to be seen in H Grey Graham, *The Social Life of Scotland in the Eighteenth Century* (1937), 532; directly concerning salters were the comments by Beveridge, *Culross and Tulliallan*, II, 245–6. More considered but equally convinced that the salters were unskilled is T C Smout, *A History of the Scottish People 1560–1830* (1972 ed), 170.
87 J R Harris, 'Skills, Coal and British Industry in the Eighteenth Century', *History*, 61 (1976), 167–182; see too A Campbell, *The Lanarkshire Miners: A Social History of their Trade Unions 1775–1874* (Edinburgh 1979), 38–45.
88 Cadell, *Rocks of West Lothian*, 321.
89 NLS, Cadell MSS, Acc 5381/Box 39/1, John Williams to William Cadell, 27 November 1776.
90 Whatley, 'Finest Place for a Lasting Colliery', 53–4.
91 They were also brought over earlier, to test the coal, see Hamilton MSS, Lennoxlove, Box 491/23/16, 'Expenses of ye Salters that went to Saltcoats to try the coall in Arran how it made salt', September 1708; Whatley, 'An Early Eighteenth Century Scottish Saltwork'.
92 NLS, Wemyss MSS, MS 7523, Tutors Sederunt Book, 1719–33.
93 NLS, Cadell MSS, Acc 5381, Bound Volume 2, Sederunt Book, Grange Colliery and Saltworks, 7 June 1779.
94 Smout, 'Henry Kalmeter's Travels', 41–2.
95 NLS, Adv MS 80.1.3, Dundas of Dundas, A Robertson to George Dundas, 20 January 1727.
96 Smout, 'Henry Kalmeter's Travels', 44.
97 Whatley, 'Finest Place for a Lasting Colliery', 68–9; Chester City Record Office, CR 352, Letter Book of Daniel Peck; *Patents of Invention* (2 vols, 1854), II, Patent 375, for 'Making & refining white salt from sea water, brine springs, and rock-salt', 5 March 1706.
98 Wemyss Castle MSS, Oncost Book, Dysart Salt, 1817–8.
99 Duckham, *Scottish Coal*, 23–5, 34–5.
100 SRO, Hope of Craighall MSS, GD 377/360, Pinkie Pans, Salt Book, 1791–9. This does appear to have been an exceptional example.
101 C A Whatley, 'The Process of Industrialisation in Ayrshire, *c*.1707–1871', Unpublished PhD thesis, University of Strathclyde, 1975, 202.
102 SRO, Henderson of Fordel MSS, GD 172/503/44, 'Description of a New Method of Manufacturing Salt by Roderick McDowall', 20 December 1813.
103 SRO, Eglinton MSS, GD3/E 60/4151, Tack, Earl of Eglinton and Hugh Currie, 1814. A similar problem faced John Taylor when he tried to erect a soda work at Newton-upon-Ayr around 1808. The Town Council refused to give permission for the works on the grounds that it would be a 'nuisance', see Whatley, 'Process of Industrialisation', 202.

CHAPTER 2

1 *RPCS*, 1st ser, II, 265–6, 442. Unfortunately, the figures have had to be drawn from the evidence of 1573 and 1574. As the salt exporting boom was clearly well under way by then and appears ultimately to have encouraged the construction of many new pans these figures probably exaggerate the industry's pre-1573 productive capacity. This is unlikely to be greatly significant though: complaints about shortages of salt within Scotland suggest that in the early stages of the upturn existing pan capacity was being stretched, with salt being diverted from home to export markets.
2 R Nicholson, *Scotland: The Later Middle Ages* (Edinburgh 1974), 439.
3 I Guy, 'The Scottish Export Trade, 1460–1599', Unpublished M Phil thesis, University of St Andrews, 1982, 158. Ms Guy's work has very recently appeared in a more accessible, though much abbreviated, form, in T C Smout (ed), *Scotland and Europe, 1200–1850* (Edinburgh 1986), 62–81.
4 Multhauf, *Neptune's Gift*, 73.
5 Guy, 'The Scottish Export Trade', in Smout, Scotland and Europe, 67.
6 Nicholson, *Scotland*, 439.
7 J N Ball, *Merchants and Merchandise: The Expansion of Trade in Europe, 1500–1630* (1977), 135–6.
8 A R Bridbury, *England and the Salt Trade in the Later Middle Ages* (Oxford 1955), 95–8.
9 Ball, *Merchants and Merchandise*, 136.
10 Bridbury, *England and the Salt Trade*, 76–93.
11 E Hughes, 'The English Salt Monopoly in the Years 1563–71', *English Historical Review*, CLVII (1925), 325.
12 See for example, J Patrick, 'The origins of the opposition to Lauderdale in the Scottish parliament of 1673', *Scottish Historical Review*, 53 (1974), 17.
13 Hughes, 'English Salt Monopoly', 335–7; J W Gough, *The Rise of the Entrepeneur* (1969), 197–9.
14 *Calendar of Scottish Papers*, II, 100; Gough, *Rise of the Entrepreneur*, 197.
15 *RPCS*, 1st ser, XIV, Addenda, 249.
16 K R Andrews, *Trade, Plunder and Settlement: maritime enterprise and the genesis of the British Empire, 1480–1630* (Cambridge 1984), 284; S G E Lythe, 'Scottish Trade with the Baltic, 1550–1650', in J K Eastham (ed), *Economic Essays in Commemoration of the Dundee School of Economics* (Dundee 1955), 66.
17 G Parker, *Spain and the Netherlands*, 1559–1659 (1979), 197.
18 Hughes, 'English Salt Monopoly', 349–50.
19 Parker, *Spain and the Netherlands*, 197.
20 F Mauro and G Parker, 'Portugal', in C Wilson and G Parker, *An Introduction to the Sources of European Economic History 1500–1800* (1977), 65–6.
21 Parker, *Spain and the Netherlands*, 197.
22 W Beveridge, *Prices and Wages in England: from the Twelfth to the Nineteenth Century* (1939), 37; similar circumstances had accounted for the sharp increase of Scottish salt exports in the 1480s. In 1485 the price of a last of Bay salt in Danzig had risen six-fold to 40 marks, whereas Scots salt was only 22 marks. See S G E Lythe, 'Economic Life', in J M Brown (ed), *Scottish Society in the Fifteenth Century* (1977), 78.
23 The course of Scottish salt exports from 1460 until 1600 is graphically illustrated in Guy, 'Scottish Export Trade', 81. I am grateful to Ms Guy for letting me see her tables and graphs prior to the completion of her thesis.

24 Lythe, 'Scottish Trade', 74–6; Guy, 'Scottish Export Trade'(thesis), 158–9.
25 S G E Lythe, *The Economy of Scotland in its European Setting 1550–1625* (Edinburgh and London 1960), 216–7.
26 R H Tawney and E Power (eds), *Tudor Economic Documents* (3 vols, 1924), II, 257–9.
27 Calculation based on figures from Lythe, 'Scottish Trade', 74; and Mauro and Parker, 'Portugal', 65.
28 SRO, Henderson of Fordel MSS, GD 172/503/34, 'Minute of Meeting of the Proprietors of the Salt Works on the East Coast of Scotland, 19 February 1794'.
29 G R Hewitt, *Scotland Under Morton* (Edinburgh 1982), 158.
30 Lythe, *Economy of Scotland*, 147.
31 *RPCS*, 1st ser, II, 293.
32 Lythe, *Economy of Scotland*, 80–3; Hewitt, *Scotland Under Morton*, 143–59; *RPCS*, 1st ser, 293, II, 295.
33 Guy, 'Scottish Export Trade', 67.
34 Lythe, *Economy of Scotland*, 84.
35 *RPCS*, 1st ser, II, 425.
36 See Bowman, 'Culross Colliery', 353–72; Cochrane, *Description of the Estate and Abbey of Culross*, 15.
37 *RPCS*, 1st ser, II, 442.
38 *RPCS*, 1st ser, III, 639.
39 *RPCS*, 1st ser, V, 51.
40 *Acts of the Parliament of Scotland*, III, 494–5.
41 *RPCS*, 1st ser, IV, 319–20; Lythe, *Economy of Scotland*, 50, 53.
42 *RPCS*, 1st ser, XII, 620.
43 *RPCS*, 2nd ser, IV, xvi, 29; Udward was involved in several types of industrial enterprise, see Lythe, *Economy of Scotland*, 36.
44 W Stephen, *The Story of Inverkeithing and Rosyth* (Edinburgh 1938), 91; A S Cunningham, *Inverkeithing, North Queensferry, Rosyth and the Naval Base* (Dunfermline 1903), 207; W Fraser, *The Sutherland Book* (Edinburgh 1894), 174; S B Calder, 'The Industrial Archaeology of Sutherland', Unpublished M Litt thesis, University of Strathclyde, 1974, 2 vols, I, 123.
45 Lythe, 'Scottish Trade', 74.
46 Lythe, *Economy of Scotland*, 217.
47 Beveridge, *Culross*, I, 149.
48 Lythe, *Economy of Scotland*, 47–8; A S Cunningham, *Romantic Culross, Torryburn, Carnock, Cairneyhill, Saline and Pitfirrane* (Dunfermline 1902), 35.
49 P Hume Brown (ed), *Early Travellers in Scotland* (Edinburgh 1973 ed), 148; Lythe, *Economy of Scotland* , 51.
50 Smout, *Scottish Trade*, 229.
51 Parker, *Spain and the Netherlands*, 37, 197.
52 See J I Israel, *The Dutch Republic and the Hispanic World 1606–1661* (Oxford 1982), 86–95, 109–17, 136, 214–6, 338–9.
53 Wemyss Castle MSS, Diary of David, 2nd Earl of Wemyss.
54 *RPCS*, 2nd ser, IV, 41.
55 Israel, *Dutch Republic*, 214.
56 Adams, 'Salt Industry of the Forth Basin', 159; Whatley, 'A Saltwork and the Community', 45.
57 Brand, *Brief Description of Orkney*, 36–7.
58 Lythe, 'Scottish Trade', 74.
59 Israel, *Dutch Republic*, 339.

60 SRO, Biel Muniments, GD 6/1252/4, 'Answers by Gosford to Dirleton's claim with Gosford's claim from Dirleton', 1698; R Mercer, 'The Archaeological Survey of a Coastal Area of East Lothian at Torness, Innerwick', *Transactions of the East Lothian Antiquarian and Field Naturalists' Society*, XV (1976), 1–23.
61 W Stevenson (ed), *The Presbytrie Booke of Kirkcaldie* (Kirkcaldy 1900), 240.
62 Beveridge, *Culross*, I, 242.
63 For example, *RPCS*, I, 3rd ser, 317; understandably, the saltmasters do appear to have been shocked by the imposition of an import tax of 16*s*. 8*d*. on Scottish salt entering Ireland and England. How soon and how far it affected them is another matter.
64 Wemyss Castle MSS, Diary of 2nd Earl.
65 Stephen, *Story of Inverkeithing*, 92.
66 Shaw, *The Northern and Western Islands of Scotland*, 160–1.
67 Whatley, 'An Early Eighteenth Century Scottish Saltwork', 92.
68 Smout, *Scottish Trade*, 229.
69 Lythe, 'Scottish Trade', 74–6.
70 Smout, *Scottish Trade*, 230.
71 NLS, Yester MSS, MS 14493, f 84, 'Proposalls concerning Salt', *c*.1670.
72 D Woodward, 'Anglo-Scottish commercial policy during the 1660s', *Scottish Historical Review*, LVI (1977), 166; the figure for the 1620s, 575 weys, from Lythe, *Economy of Scotland*, 217, relates to London alone, and may therefore underestimate the quantity of salt being shipped south.
73 Public Record Office, SP 29/233, 241, 243; these papers, produced on behalf of the Shields salt industry during 1667, purport to demonstrate, amongst other things, that Scottish production costs were around one-third below those of north east England. How accurate the respective figures are must be open to doubt. For instance the Scottish costs do not include any sums for the annual binding money which was clearly paid to the salters. There are also other criticisms of the methods used to calculate the figures. For a fuller account see E Hughes, 'The negotiations for a Commercial Treaty between England and Scotland in 1668', *Scottish Historical Review*, 24 (1927), 37–8. For a discussion of payments made to salters, see Chapter Five.
74 Woodward, 'Anglo-Scottish commercial policy', 174.
75 Smout, *Scottish Trade*, 230.
76 T C Barker, 'Lancashire Coal, Cheshire Salt and the Rise of Liverpool', *Transactions of the Historic Society of Lancashire and Cheshire*, CIII (1951), 92.
77 A Mitchell (ed), *MacFarlane's Geographical Collections* (Edinburgh 1907), II, 408.
78 Cochrane, *Description*, 15.
79 Tucker, 'Report', 5.
80 Lythe, *Economy of Scotland*, 176.
81 NLS, Yester MSS, MS 14493, f 84; *Remarks Upon a Pamphlet, Intitled, The Considerations in relation to Trade Considered, and a Short View of our present Trade and Taxes Reviewed* (Edinburgh 1706), 8.
82 SRO, Salt Charge Vouchers, E 536/3–5.
83 Smout, *Scottish Trade*, 230.
84 SRO, Hamilton MSS, GD 406/1/4137, Daniel Hamilton to Earl of Arran, 18 May 1696.
85 Ellis, 'Tyneside Salt Industry', 49.
86 R Mitchison, *Lordship to Patronage: Scotland 1603–1745* (1983), 109–10.
87 I Bisset, 'Commercial and Economic Developments in Late 17th Century

Dundee', Unpublished MA dissertation, Department of Modern History, University of Dundee 1978, 14; for the buoyancy of the coastal salt (and coal) trade into Dundee at this time see Dundee Archive Centre, DARC, TC/TS/3–4, Dundee Register of Ships, 1701–13.

88 Whatley, *'That Important and Necessary Article'*, 30.

89 D R Hainsworth (ed), *The Correspondence of Sir John Lowther of Whitehaven 1693–1698* (Oxford 1983), 538.

90 See Chapter Four.

91 J Butt and I L Donnachie, *The Industrial Archaeology of the British Isles* (London 1979), 138.

92 D Turnock, *The Historical Geography of Scotland Since 1707* (Cambridge 1982), 58; B Lenman, *From Esk to Tweed* (Glasgow 1975), 39; W H Marwick, *A Short History of Labour in Scotland* (Edinburgh 1967), 1; for an older version of this interpretation, see R Sibbald, *The History of Fife and Kinross* (Cupar 1800 ed), 340.

93 Lenman, *Economic History*, 125.

94 NLS, Adv MS 80.1.4, George Beattie to Robert Dundas, 5 and 15 February 1740; A Cochrane, *Thoughts on the Manufacture and Trade of Salt* (1794), 6; J Girvin, *The Impolicy of Prohibiting the Exportation of Rock salt from England to Scotland* (1799), 32–3; SRO, Customs and Excise Records, CE 67/1/36, Salt Letter Books, Collector (Alloa) to the Board, 6 May 1796; *Old Statistical Account of Scotland*, XVI (1795), Whithorn, 292.

95 On 20 August 1772 for example the partners of Grange colliery and saltworks noted that 'the sale of Salt from the Girnels Since the Commencement of the works to this date falls short of the Quantity put therein—858 Bushels'. NLS, Cadell MSS, Acc 5381, Bound Volume 2, Memorandum Book, Grange Colliery, 1771–1791; see too, Whatley, 'A Saltwork and the Community', 54.

96 It should also be noted that data from many thousands of salt vouchers, some of which were in a very poor condition, had to be transcribed and therefore it is possible that some (hopefully minor) errors have occurred during this process. As perhaps as much as one-seventh of the salt actually may have wasted in the girnels during periods of slack demand the quantity of salt sold was almost certainly less than that which was made at the pans. Largely for these reasons, in the text, numbers of bushels have been rounded to the nearest whole and discussion has usually been conducted in terms of 100s or 1,000s of bushels.

97 SRO, Ailsa MSS, GD 25/8/6/867, 'Tack: Sir Archibald Kennedy of Culzean to John Kennedy yr thereof & Hugh Hamilton, mercht, Ayr, of the Saltpan of the maiden head of Turnberry', 5 August 1709.

98 For a useful if rather uncritical account of the activities of Duchess Anne, see R K Marshall, *The Days of Duchess Anne: Life in the Household of the Duchess of Hamilton 1656–1716* (1973); for specific material relating to the beginnings of 'industrial' development on Arran, see correspondence between several members of the Hamilton family: SRO, Hamilton MSS, GD 406/1/4877, 6557, 6597.

99 See note 67 above.

100 E Hughes, *Studies in Administration and Finance 1558–1825* (Manchester 1934), 413.

101 *Calendar of Treasury Papers*, CCXLVII, 275.

102 Quoted in Hughes, *Studies in Administration*, 413.

103 Apparently too, great quantities of foreign salt were imported just prior to the Union. While this salt was supposed to be re-exported, it is posssible that some

of it was sold to domestic consumers: see R Renwick (ed), *Extracts from the Burgh Records of Glasgow 1691–1717* (Scottish Burgh Record Society 1908), 456.

104 SRO, GD1/170/714/3/33373, York Buildings Society Papers, correspondence between John Cockburn and John Horsley.

105 SRO, E 536/84, Salt Charge Vouchers, Montrose, 1797–8.

106 Girvin, *Impolicy*, 15.

107 NLS, Yester MSS, MS 14493, f 84.

108 See T C Smout, 'The Development and Enterprise of Glasgow 1556–1707', *Scottish Journal of Political Economy*, 7 (1960), 203; R H Campbell, 'The Anglo-Scottish Union of 1707. II. The Economic Consequences', *Economic History Review*, XVI (1964); C Gulvin, 'The Union and the Scottish Woollen Industry 1707–1760', *Scottish Historical Review*, 50 (1971); A J Durie, *The Scottish Linen Industry in the Eighteenth Century* (Edinburgh 1979), 9.

109 Hughes, *Studies in Administration*, 414; Clows, *Chemical Revolution*, 58.

110 *The History and Proceedings of the House of Commons* (1742), IV, 19.

111 Ellis, 'Tyneside Salt Industry', 49–50; 'A Bold Adventurer: The Business Fortunes of William Cotesworth *c.*1668–1726', *Northern History*, XVII (1981), 121–2. Earlier writers took a very different view from that of either Hughes or the Clows. In Sibbald's *History of Fife*, for instance, it is argued that when duties were imposed on Scottish salt at the Union, the industry 'could not maintain the competition with English mineral salt'. C Tomlinson, *The Natural History of Common Salt* (1850), 267, was of the opinion that equalisation of English and Scottish salt duties in 1707 'greatly injured' the Scottish salt industry. Where both of these writers appear to have gone wrong was in assuming that the duty arrangements seriously damaged the Scottish industry.

112 Public Record Office, T 64/232/757, Report (Customs) on Berwick District, 20 September 1733; SRO, E 536/19, Salt Charge Vouchers, 1733–4.

113 Ellis, 'Tyneside Salt Industry', 51, 56; for a recent account of English antipathy towards the Scots in the eighteenth century, see M Duffy, *The Englishman and the Foreigner* (Cambridge 1986), 18–20; I am grateful to Joyce Ellis of Loughborough University for providing me with information about Tyneside fears. For a useful account of the Lymington indusry, see D A E Cross, 'The Salt Industry of Lymington', *Journal of Industrial Archaeology*, 2 (1965).

114 Smout, *Scottish Trade*, 230; and see, for example, Wemyss Castle MSS, Account Book, Dysart Saltworks, 1697–1715. Notably, the firmest claim about Scottish dependence upon salt sales in England relates to the early nineteenth century. It was alleged that during the Napoleonic Wars Joppa pans were 'almost wholly' dependent upon contraband trade, Baird, *Annals of Duddingston*, 25. If this claim is accurate, it is probably to be explained by war-induced shortages of salt in the north east of England, made more acute by the earlier demise of the Tyneside salt industry.

115 See, for example, M Gray, *The Highland Economy 1750–1850* (Edinburgh 1957), 112–3; *The Fishing Industry of Scotland 1790-1914: A study in Regional Adaptation* (Oxford 1978), 6.

116 *Scots Magazine*, 66 (1804), 761–2.

117 By the late eighteenth century the scale of legal English imports had grown considerably. In the five years between 1793 and 1797, on average, 345,000 bushels of English salt entered Scotland. At the same time the figure for annual average recorded sales of Scottish salt was just under 300,000 bushels. See Girvin, *Impolicy*, 15; SRO, E 536/80–4, Salt Charge Vouchers, 1793–8.

118 Barker, 'Lancashire Coal, Cheshire Salt', 92–4; W H Chaloner, 'Salt in Cheshire, 1600–1870', *Transactions of the Lancashire and Cheshire Antiquarian Society*, 72 (1961), 68–70.
119 W T McIntyre, 'The Salt Pans of the Solway', *Transactions of the Cumberland and Westmorland Antiquarian Society*, XLII (1942), 12.
120 Edinburgh University Library, Laing MSS, Add 4/16, 'Reasons for allowing English Salt to be brought to Scotland at the Scotch Duty'; *Scots Magazine*, 66(1804), 763; Girvin, *Impolicy*, 40–1; Cochrane, *Present State*, 9.
121 NLS, Adv MS 80.1.4, George Beattie to Robert Dundas, 5 February 1740.
122 *Scots Magazine*, 66 (1804), 821. Although this invaluable article was published in 1804, it appears to have been written in 1787.
123 *Ibid.*
124 SRO, Hope of Craighall MSS, GD 377/360, Pinkie Pans Salt Book, 1791–9.
125 For a more detailed version of this argument, see C A Whatley, 'Sales of Scottish Marine Salt, *c.*1713–1823', *Scottish Economic and Social History*, 6 (1986), 10–11.
126 See S Parkes, *Chemical Essays* (1815), IV, essay on 'Bleaching'; J R Hume, 'The St Rollox Chemical Works 1799–1964', *Industrial Archaeology*, 3 (1966), 188; *Journals of the House of Commons*, 37 (1780), 893–6; Girvin, *Impolicy*, 23; *Parliamentary Papers*, XIV(1817), Papers relating to the Salt Duties, 388–414.
127 E Gauldie, 'Scottish Bleachfields 1718–1862', Unpublished B Phil thesis, University of St Andrews, 1966, 334–5; Durie, *Linen Industry*, 104.
128 Calculated from SRO, E551/17, Excise Revenue Accounts, 1821–2.
129 G Hay, *History of Arbroath* (Arbroath 1899), 375; J M McBain, *Arbroath: Past and Present* (Arbroath 1877), 139; Calder, 'Industrial Archaeology of Sutherland', I, 135.
130 SRO, Eglinton MSS, GD 3/E 60/4151, Tack, the Earl of Eglinton and Hugh Currie, 1814; SRO, Coal Board Records, CB 27/3, Robert Bald's Report on Bartonholm, Shewalton and Doura collieries, 20 January 1818; Ayr Advertiser, 8 November 1821.
131 Whatley, 'Sales of Marine Salt', 11.
132 *Ibid*, 10.
133 Whatley, *'That Important and Necessary Article*', 35.
134 See SRO, E 536, Salt Charge Vouchers. The first vouchers are available from 1713.
135 A S Cunningham, *Dysart: Past and Present* (Edinburgh 1912), 40; SRO, Rosslyn MSS, GD 164/45/348, Accounts of Dysart Coal and Salt, 1820– 6; SRO, Henderson of Fordel MSS, GD 172/862, Fordell Colliery Returns, 1823–43; see also SRO, Hope of Craighall MSS, GD 377/140/365, Accounts of Craighall Colliery and Pinkie Pans, 1812–62.
136 See, for example, Salmon, *Borrowstoness*, 161–2.
137 SRO, Henderson of Fordel MSS, GD 172/1107/34, Salt Agent's copy letter book, 14 November 1846.
138 See, for example, *New Statistical Account*, IX, 187; NLS, Cadell MSS, Acc 5381, Box 39/I, 'Inventory of Plant connected with Cockenzie Saltworks as at March 1870'; Baird, *Annals of Duddingston*, 27.
139 SRO, Hope of Craighall MSS, GD 377/130/1, 'Notice of Sale of Pinkie Salt Works', 4 October 1855.
140 See Salmon, *Borrowstoness*, 162; SRO, Dissolved Companies Files, BT 2/1888/1, Papers of the Scottish Salt Company Limited; Baird, Annals of Duddingston, 27; for other post-1850 developments see Whatley, '*That Important and Necessary Article*', 42.

141 Calculations based upon SRO, E 536/3–5, Salt Charge Vouchers, 1716–9.
142 For a discussion of population distribution in Scotland in the 1690s, upon which this comment is based, see Flinn, *Scottish Population History*, 187–99.
143 Tucker, 'Report', 5; J Kay, *A Series of Original Portraits* (Edinburgh 1872 ed), II, 166; *Old Statistical Account* (1976 ed), II, The Lothians, 229, 293–4.
144 Kay, *Original Portraits*, 166.
145 Edinburgh University Library, Laing MSS, Add 4/16, 'Reasons for allowing English salt...at the Scotch Duty'.
146 SRO, Customs and Excise Records, CE 67/1/36, Salt Letter Books, Collector(Alloa) to the Board, 1786–96, 6 May 1796.
147 See, for example, NLS, Cadell MSS, Acc 5381, Box 39/I, James Morison to W Cadell, 30 June 1777; 'Petition of the Manufacturers of Salt in that county [Ayrshire]' and 'Declaration by the J P's and Freeholders of the county of Ayr'(nd); SRO, Leven and Melville MSS, GD 51/5/190, Robert Hanna to Henry Dundas, nd(*c*.1795). The problem became worse as the level of duties rose, see Girvin, *Impolicy*, 33.
148 See note 146 above.
149 *Scots Magazine*, 66 (1804), 760.
150 SRO, CS 237/1 P/M/3/1, S Mitchell v W Orr, 1764, Miscellaneous Accounts, 1756–9.
151 Wemyss Castle MSS, Oncost Book, Dysart Salt, 3 May 1817–7 February 1818.
152 SRO, Rosslyn MSS, GD 164/49/398, Coal and Salt Accounts, 1714–7.
153 SRO, Ailsa MSS, GD 25/9/Box 46, 'Book of Accompts of Salt Sould in ye yr 1757'.
154 SRO, CS 237/1 P/M/3/1, Mitchell v Orr, Miscellaneous Accounts.
155 NLS, Wemyss Castle MSS, MS 10686, John Chalmers to John Rose, 10 December 1800.
156 Calder, 'Industrial Archaeology of Sutherland', I, 123.
157 Whatley, *'That Important and Necessary Article'*, 35.
158 J Roger, *General View of the Agriculture of the County of Angus or Forfar* (Edinburgh 1794), 16; India Office Records, H/730, 'Answers to Mr Scotts Queries the 14th June 1796'; Calder, 'Industrial Archaeology of Sutherland', I, 137; R J Adam (ed), *Papers on Sutherland Estate Management 1802–1816* (1972), II, 222, 254.
159 L B Taylor (ed), *Aberdeen Council Letters* (Oxford 1961), III, 276.
160 Wemyss Castle MSS, Ledger of Salt Accounts, 1771–6.
161 *Ibid*, William Wilson to John Grieve, 19 August 1771.
162 NLS, Cadell MSS, Acc 5381, 39/1, William Alexander to William Cadell, 21 and 26 May 1777.
163 SRO, Ailsa MSS, GD 25/9/398, 'Book of Accompts'.
164 SRO, CS 237/1 P/M/3/1, Mitchell v Orr, Miscellaneous Accounts; *Old Statistical Account*, VII (1793), 19. It went inland as far to the north and east as Renfrewshire and 'Clydesdale', until it met 'the Borrowstoness salt'.
165 Whatley, 'Scottish salt making', 8.
166 Dr Singer, *General View*, 530; the level of hostility to the use of Scottish salt for commercial curing and preserving purposes in Galloway and elsewhere in Scotland can be see in the following: *Scots Magazine*, 55(1793), 496; SRO, Leven and Melville MSS, GD 51/5/201, Patrick Heron to Henry Dundas, 23 September 1793; Broughton and Cally MSS, GD 10/1421/Vol 7/352, 'Memorial from JPs and Commissioners of Supply for Stewartry of Kirkcudbright', *c*.1787.
167 Edinburgh University Library, Laing MSS, Add 4/16, 'Reasons'.

CHAPTER 3

1 Lenman, *Economic History*, 124–5; Adams, 'Salt Industry of the Forth Basin', 161.
2 Guy, 'Scottish Export Trade', 67.
3 Duckham, *Scottish Coal*, 15.
4 M H B Sanderson, 'The Mauchline Account Books of Melrose Abbey, 1527–1528', *Ayrshire Collections*, 11 (1973–6), 92; T C Smout and A Fenton, 'Scottish Agriculture before the Improvers: an Exploration', *Agricultural History Review*, XIII(1965), 82–3.
5 See B C Skinner, *The Lime Industry of the Lothians* (Edinburgh 1969); for a local study of the impact of lime burning on coal, see Whatley, 'The Finest Place for a Lasting Colliery', 64.
6 Duckham, *Scottish Coal*, 24–8.
7 Cadell, *Rocks of West Lothian*, 321.
8 NLS, Yester MSS, MS 7033, 'Copy of the Saltowners letter to me [Marquess of Tweeddale]', 3 January 1668.
9 T Keith, 'Economic Condition of Scotland under the Commonwealth and the Protectorate', *Scottish Historical Review*, V(1908), 281; J U Nef, *The Rise of the British Coal Industry* (2 vols, 1932), II, 228–30.
10 SRO, Mar and Kellie MSS, GD 124/17/522, 'Information anent great coalls', 1683; Lenman, *Economic History*, 40.
11 Nef, Coal Industry, I, 118–9.
12 SRO, Hamilton MSS, GD 406/C1/4292, Daniel Hamilton to [Earl of Arran ?], 18 January 1699; earlier, in May 1698, the Earl of Arran had been informed by the same source that the principal reasons why 'no great profit' had been derived for 17 months from the estate coalwork were, first, that 'severalls of the Salt pans have been to repair', and secondly, that 'the great floods of rain which we have had both in harvest and this winter did for some tyme both set the pans Idle and Drounded the coal'.
13 SRO, Hamilton MSS, GD 406/C1/4137, Daniel Hamilton to Duke of Hamilton, 3 February 1702.
14 SRO, Hamilton MSS, GD 406/1/4975, 'Memorandum concerning Coall and Salt', nd, c.1702.
15 *Remarks for the Salt-Masters, Fishers of Salmond, Herrings and White-Fish, and others who make use of Scots Salt, Humbly offered upon the Eighth Article of the Treaty on Union* (Edinburgh 1707).
16 SRO, Hamilton MSS, GD 406/C1/4974, Daniel Hamilton to Duke of Hamilton, 23 April 1702.
17 Calculated from SRO, CS 96/4460, Grange Coal Accounts, 1695– 1702.
18 Nef, *Coal Industry*, I, 48–9.
19 Calculated from SRO, RH9/1/33, Accounts of Kincardine and Tulliallan Coal and Salt Works, 1679–80.
20 Smout, *Scottish Trade*, 227.
21 See Duckham, *Scottish Coal*, 19–20; C A Whatley, review of M W Flinn, *The History of the British Coal Industry* (Oxford 1984), *Scottish Historical Review*, LXIV (1985), 198.
22 Nef's figure of 150,000 tons is a very crude estimate, based largely upon the assumption that at the end of the seventeenth century the Scottish salt industry was bigger than that of Tyneside, *Coal Industry*, I, 207–8. He also assumed that Tulliallan colliery, which turned out around 15,000 tons of coal, most of which

was used in the saltpans, was a 'fair sample' of 'the more important Scottish collieries', I, 49. (It was in fact rather a large colliery). As a result he concluded that the Scottish salt industry must have burned more than the 125,000 tons reckoned to have been used in Durham and Northumberland saltworks. Yet it has recently become quite clear that the salt industry of the north east of England was substantially bigger than its Scottish rival, producing between 8–9,000 tons around 1605 and 12–13,000 tons in 1630. By the end of the century the 170 or so pans at Shields were capable of producing 25,000 tons per annum. See S M Jack, *Trade and Industry in Tudor and Stuart England* (1977), 87–9; Ellis, 'Tyneside Salt Industry', 50. The Scottish industry had nowhere near this productive capacity. A contemporary estimate of salt production in 1670 was 83,000 large bolls, or 332,000 bushels (NLS, Yester MSS, MS 14493, f 90, 'Proposalls concerning salt', *c*.1670). This is equivalent to 8,300 tons. To make this quantity of salt would have required 49,800 tons of coal. In the early eighteenth century the highest contemporary estimate for Scottish salt production was 464,944 bolls. Actual annual average recorded output for the 1710s was under 250,000 bushels. Hence the quantity of coal consumed in Scottish pans was between 37,500 and 74,226 tons. The actual figure was probably nearer to the lower of these. Owing to his error of assuming a coal-salt ratio of 16:1 tons, Duckham's figure for saltpans consumption of coal in the period 1771–83, of 140,000 tons, is also far too high (*Scottish Coal*, 26). His estimated annual output of just over 349,500 bushels of salt would have burned just over 52,000 tons of small coal.

23 Duckham, *Scottish Coal*, 18.

24 Ibid, 230; this, admittedly, is the lowest figure in the series. The average annual quantity exported from the Forth during the 1790s was 25,776 tons.

25 Whatley, 'Finest Place for a Lasting Colliery', 58–61.

26 Duckham, *Scottish Coal*, 198–9; for a recent assessment of the British coal industry's profitability, which partly relies upon Scottish evidence, see M W Flinn, *British Coal Industry*, 326.

27 R Bald, *A General View of the Coal Trade of Scotland* (Edinburgh 1808), 25; NLS, Tweeddale MSS, MS 14551, f 61, William Adam to Marquess of Tweeddale, 4 June 1743; f 141, John Adam to Marquess of Tweeddale, 7 March 1758; f 147, John Adam to George Hay, 3 December 1764.

28 This is suggested in Table 11.1, in Flinn, *British Coal Industry*, 388–9; see too Whatley, 'Finest Place for a Lasting Colliery', 85–8; SRO, CS 238/S/10/9, George Glasgow to Walter Scott, 21 February 1774.

29 C A Whatley, 'From Servitude to Subjugation? Legal Records, Labour Relations and Collier "Emancipation" in the Eighteenth Century', *Scottish Records Association Conference Report*, 5 (1986), 15.

30 SRO, Dundonald MSS, GD 233/109/5/126, Archibald Cochrane to Samuel Mitchelson, 10 March 1777.

31 SRO, Mar and Kellie MSS, GD 124/17/529, 'Memorandum re chows, Alloa 5 August 1703'.

32 By the early eighteenth century few pans remained in operation on the upper reaches of the Forth west of Culross. Little salt was made at Kennetpans and in 1716–7, for instance, only 137 bushels were sold from the Clackmannan pans. SRO, E 536/3, Salt Charge Vouchers, 1716–7.

33 NLS, Pitfirrane MSS, MS 6482, f 6, 'Copy Ratification in favour of Dame Janet Halkett', 21 March 1707; f 9, 'Anent the Memorial given in...by Sir Peter Halkett of Pitfirrane', 8 May 1707.

34 R Douglas, 'Coal-mining in Fife in the second half of the eighteenth century', in G W S Barrow, *The Scottish Tradition* (Edinburgh 1974), 216.
35 Cunninghame District Council Archives, Auchenharvie MSS, Misc, 'Gen'll Ballce, 1749–61'.
36 SRO, Fairholme MSS, GD 237/143/2, 'Memorial—Dr John Roebuck to his Creditors', 1771.
37 NLS, Cadell MSS, Acc 5381, Bound Volume 2, 15 November 1779.
38 P L Payne, 'The Halbeath Colliery and Saltworks, 1785–1791', in A Slaven and D H Aldcroft (eds), *Business, Banking and Urban History* (Edinburgh 1982), 17.
39 SRO, Coal Board Records, CB 27/2, Robert Bald's Report on the Coal and Saltworks of Dysart, 13 November 1817.
40 Duckham, *Scottish Coal*, 233; Adam, *Sutherland Estate Management*, II, 235.
41 SRO, Henderson of Fordel MSS, GD 172/835/2, 'General State or Abstract of Fordel Books for the year 1791'.
42 SRO, Rosslyn MSS, GD 164/45/348, 'Account of Dysart Coal and Salt, February 1818–May 1820'. How the value of coal had increased, and how far the gap between the value of coal and salt had widened, can be seen by comparing these figures with those from 1792, when coal income was £574 and that from salt £201 (GD 164/45/337).
43 SRO, CS 96/3086, P Warner v R R Cunninghame, 1799, 'Supplementary Report of John Clerk of Eldin and Dr John Roebuck upon the Coalworks of Ardeer and Saltworks at Saltcoats'.
44 See note 41 above.
45 NLS, Cadell MSS, Acc 5281, 39/1, 'State of Maryburgh and Craigie Saltpans', 4 March 1777; James Morison to William Cadell, 30 June 1777.
46 T C Smout, 'Scottish landowners and economic growth, 1650–1850', *Scottish Journal of Political Economy*, XI(1964), 218–34.
47 G Donaldson (ed), *Protocol Book of James Young 1485–1515* (Edinburgh 1952), 103.
48 SRO, CS 229/P1/30, List of Feuars, Prestongrange estate, c. 1744.
49 SRO, E 536/3, Salt Charge Vouchers, 1716–7.
50 Lenman, *Economic History*, 41–3; Durie, *Linen Industry*, 14, 55, 72; 'Lairds, Improvement, Banking and Industry in Eighteenth Century Scotland: Capital and Development in a backward economy—a case study', in T M Devine (ed), *Lairds and Improvement in the Scotland of the Enlightenment* (Dundee 1978), 22.
51 Duckham, *Scottish Coal*, 14.
52 Wemyss Castle MSS, Diary of Earl David.
53 The greatest part of the expense comprised the pan plates and other iron components which were generally made from expensive high-grade Swedish iron. This requirement became less pressing in the latter decades of the eighteenth century, after Carron Company began making pan plates.
54 SRO, Henderson of Fordel MSS, GD 172/503/2, James Pinkerton to Sir John Henderson, 28 March 1788.
55 *Scots Magazine*, 66(1804), 760.
56 Lythe and Butt, *Economic History*, 187.
57 SRO, Callendar of Crichton MSS, GD 247/Box 84/2/11, 'Valuation of wrought & cast iron work of Salt pans of Alexander Paterson, leased from Sir John Callendar', 16 February 1813.
58 Pans were also more often publicly advertised for lease or sale. see, for example, *The Fifeshire Advertiser*, 1 January 1853.
59 SRO, Rosslyn MSS, GD 164/Box 76/802/6, 'Offer, of the Saltworks at Dysart, belonging to Sir James St Clair Erskine of Sinclair...1797'.

60 Wemyss Castle MSS, Diary of Earl David, 25 August 1634.
61 A S Cunningham, *Randolph Gordon Erskine Wemyss: An Appreciation* (Edinburgh 1910), 34.
62 Calculated from SRO, CS 96/4520, Factor's Accounts, Prestongrange, 1716–23.
63 I Whyte, *Agriculture and Society in Seventeenth Century Scotland* (Edinburgh 1979), 33–6.
64 Smout, *Scottish Trade*, 125–8.
65 Mitchison, *Lordship to Patronage*, 94–5.
66 SRO, Hamilton MSS, GD 406/C1/5189, Daniel Hamilton to Duke of Hamilton, 29 September 1704.
67 The Hamilton MSS during this period provide a vivid impression of the economic difficulties of the 1690s and early 1700s.
68 SRO, Rosslyn MSS, GD 164/49/398/10, 'Acctt of Money Depursed...The Lord St Clair's Salt Works Since 15th July 1713'.
69 For labour shortages in coal mining, see B F Duckham, 'Serfdom in Eighteenth Century Scotland', *History*, LIV(1969), 178, 193–4; for an example of what occurred when colliers were not paid promptly, see Hamilton MSS, Lennoxlove, M10/52, 'Memorandum for Lord Basil Hamilton, 17 June 1695'. At the time of writing part of the Hamilton MSS was being transferred to the SRO and relisted. Where a new reference number was not available, the Lennoxlove reference has been given. A full list of all the Hamilton MSS is held in the SRO.
70 NLS, Adv MS 80.1.3, Dundas of Dundas, Thomas Hope to [Tutors of Wemyss of Bogie estate], 28 June 1722.
71 SRO, Elphinstone MSS, GD 156/Box 11/6/3, Accounts, Elphinstone Pans, 1716–9.
72 Hamilton Public Library, Hamilton MSS, Bundle 4, 'Accompt of Salt to his Grf the Duke of Hamilton's family', 1690–3; SRO, Ailsa MSS, GD 25/8/6/867, Tack, 5 August 1709.
73 SRO, Agnew of Lochnaw MSS, GD 154/461/4, Tack, Sir Andrew Agnew and Patrick McNeilly, 24 May 1762.
74 SRO, Eglinton MSS, GD3/E 127/5758, Tack, Earl of Eglinton and William Bone, 1778.
75 SRO, Henderson of Fordel MSS, GD 172/442/1, Tack, Sir Robert Henderson and Hugh Wilkie, 1746.
76 See Duckham, *Scottish Coal*, 141–99; Whatley, 'Process of Industrialisation', 66–85.
77 SRO, Ailsa MSS, GD 25/8/6/867, Tack, 5 August 1709.
78 SRO, Ailsa MSS, GD 25/9/Box 46, 'Inventory of the pann of Turnberry', 1728.
79 NLS, Fletcher of Saltoun MSS, MS 17680, 'Memorial and proposal to his Grace the Duke of Argyle by Samuel and William Mitchell, 1759'; see too, Whatley, 'Scottish salt making', 10–11.
80 Signet Library, CSP 59, 'Petition of Francis Holburn, provost of Inverkeithing and others, 1761'.
81 For a recent account of the Ayr Bank, see C W Munn, *The Scottish Provincial Banking Companies 1747–1864* (Edinburgh 1981), 29–36.
82 Records of the Burgh of Prestwick (Glasgow 1834), 95–6.
83 Whatley *'That Important and Necessary Article'*, 36–7, 68.
84 SRO, Rosslyn MSS, GD 164/Box 76/802/6, 'Offer...1797'.
85 Whatley, *'That Important and Necessary Article'*, 39.
86 SRO, Eglinton MSS, GD 3/E 60/4151, Tack, Earl of Eglinton and Hugh Currie, 1814.

87 Unfortunately, the nineteenth century coal mining industry in Scotland still awaits its historian. For excellent accounts of developments in ownership and management in two of the mining districts see Campbell, *Lanarkshire Miners*, 93–116 and J A Hassan, 'The landed estate, paternalism and the coal industry in Midlothian, 1800–1880', *Scottish Historical Review*, LIX(1980), 73–91.

CHAPTER 4

1 A Smith, *An Enquiry into the Nature and Causes of the Wealth of Nations*, (2 vols, 1960 ed), 117.
2 For a modern account of Smith's life, see R H Campbell and A S Skinner, *Adam Smith* (1982).
3 See H Hamilton, 'Combination in the West of Scotland Coal Trade, 1790–1817', *Economic History*, II(1930); Duckham, *Scottish Coal*, 235–8.
4 H Hamilton, *The Industrial Revolution in Scotland* (1966), 204–6.
5 See P M Sweezy, *Monopoly and Competition in the English Coal Trade 1550–1850* (Cambridge 1938); P Cromar, 'The Coal Industry on Tyneside 1715–1750', *Northern History*, XIV(1978); Flinn, *Coal Industry*, 254–73.
6 Mitchison, *Lordship to Patronage*, chapters 1-5.
7 Smout, 'Scottish Landowners', 229.
8 *Acts of the Lords of Council 1501–1554* (Edinburgh 1932), 271.
9 *Ibid*, 364.
10 Hughes, 'English Salt Monopoly', 348–9.
11 *RPCS*, 1st ser, II, 425.
12 *Ibid.*
13 *RPCS*, 1st ser, IV, 319.
14 *RPCS*, 1st ser, IX, 306–7.
15 *RPCS*, 2nd ser, II, xxxi, 72, 561; while it is unclear how genuine their concern was for their workers, the coal and salt masters did often include in their protests against change the plea that any harm done to their industries would also affect the livelihoods of the many thousands of people they claimed they employed. Thus later in the century the colliery and saltwork proprietors tried to resist additional taxation, partly on the grounds that additional costs would 'ruine...Many thousands of people...not only coalziers and salters themselves & their children & servants but also people of all trades and occupationsfrom whom they buy their necessaries will be extremely prejudiced', SRO, Rosslyn MSS, GD 164/49/398, 'Reasons to be given in against the valuation of coall and salt works being Casuall rents', nd.
16 *RPCS*, 2nd ser, VI, 344.
17 *Acts of the Parliament of Scotland*, VI, 447.
18 *Acts of the Parliament of Scotland*, VII, 88.
19 *RPCS*, 3rd ser, II, 53–4.
20 *Ibid.* The fishing industry was to be encouraged, but not at the expense of the salt and coal industries.
21 Mitchison, *Lordship to Patronage*, 94.
22 Taylor, *Aberdeen Council Letters*, IV, 275–6.
23 *Ibid.*
24 NLS, Yester MSS, MS 7033, 'Copy of the saltowners letter to me [Tweeddale], 3 January 1668'.
25 NLS. Yester MSS, MS 14493, f 82, 'Papers anent the Salt', 1669–70; *RPCS*, 3rd ser, III, 166, 311.

26 *Ibid*, f 84, 'Instructions to ye Salt Manadgers', 24 August 1670; 'Additional Instructions to ye Salt Managers', 10 March 1671; MS 7034, f 24, 'Copye Letter from the Exchequer anent the Salt', August 1673.
27 G Mackenzie of Rosehaugh, *Memoirs of the Affairs of Scotland from the Restoration of King Charles II* (Edinburgh 1821), 242.
28 R Mitchison, *A History of Scotland* (1982 ed), 254.
29 Mackenzie, *Memoirs*, 171, 242; see too, O Airy (ed), *The Lauderdale Papers* (1885), II, 243–7; III, 1. Kincardine became as unpopular as Sir Walter Seton had been from around 1665 until 1670, when he had been responsible for gathering all of the duties on salt in Scotland. In some of the undated memoranda of the period it is sometimes difficult to determine which individual is being referred to. In a recent article (*Scottish Historical Review* 1987) I have erroneously given Kincardine the title by which Seton was once described, 'absoluit master of all the Salt in the Kingdom'.
30 Patrick, 'Opposition to Lauderdale', 17.
31 Yester MSS, MS 7034, f 24, 'Copye Letter from the Exchequer anent the Salt', August 1673.
32 *Ibid.*
33 *Acts of the Parliament of Scotland*, VIII, 211.
34 See, for example, Mitchison, *History of Scotland*, 253–72; and more specifically, Patrick, 'Opposition to Lauderdale', 6–21.
35 *Ibid*, 19–21.
36 Airy, *Lauderdale Papers*, II, 246–7.
37 *Ibid*, III, 1; Mackenzie, *Memoirs*, 258–60.
38 *RPCS*, 3rd ser, VII, 653.
39 G P Insch, *The Company of Scotland* (1932), 24; see too Mitchison, *Lordship to Patronage*, 106–7.
40 *Acts of the Parliament of Scotland*, VIII, 356.
41 Taylor, *Aberdeen Council Letters*, VI, 343–4.
42 Mackenzie, *Memoirs*, 258.
43 Chaloner, 'Salt in Cheshire', 67; it is also worth noting that there was an overland route, which is still called 'Salter's Way', by which Tyneside salt could be brought into the Scottish border counties. Presumably this only went out of use after 1707. See J T White, *The Scottish Border and Northumberland* (1973), 163, 206.
44 For a description of the process used by the Dutch, see Ramsay, 'Culinary Salt', 379–81; for the impact of their salt making and fish preservation techniques and organisation, see R W Unger, 'Dutch Herring, Technology, and International Trade in the Seventeenth Century', *Journal of Economic History*, XL(1980), 253–79.
45 *RPCS*, 2nd ser, XII, 620.
46 *Acts of the Parliament of Scotland*, XII, 80.
47 SRO, Hamilton MSS, GD 406/C1/6847, 6871–3, 6847, 6281, correspondence between members of the Hamilton family concerning the African Company and salt manufacturing, September and October 1696.
48 *Ibid.*
49 For a brief and telling account of what is more commonly termed the 'Darien Scheme', see W Ferguson, *Scotland: 1689 to the Present* (Edinburgh 1968), 27–32.
50 *Acts of the Parliament of Scotland*, X, 293; University of London, Goldsmith's Library, 'Draught of an Act Anent Coal and Salt', 1701.

51 *Acts of the Parliament of Scotland*, XI, Appendix, 74.
52 T C Smout, 'The Road to Union', in G Holmes (ed), *Britain after the Glorious Revolution 1689–1714* (1969), 176; R H Campbell and J B Dow, *Source Book of Scottish Economic and Social History* (Oxford 1968), 105; Butt, *Industrial Archaeology*, 13; T Dickson (ed) *Scottish Capitalism: Class, State and the Nation from before the Union to the Present* (1980), 86; Mitchison, *Lordship to Patronage*, 129–35. Historians have paid relatively little attention to the concern which many Scots, including members of the landed classes, felt about the prospect of having to compete on equal terms with the English. A recent exception is R H Campbell, *The Rise and Fall of Scottish Industry 1707–1939* (Edinburgh 1980), 6–7.
53 Dickson, *Scottish Capitalism*, 86; P W J Riley, *The Union of England and Scotland: A Study in Anglo-Scottish politics of the eighteenth century* (Manchester 1978), 201.
54 Smout, *Scottish Trade*, 231.
55 *Historical Manuscripts Commission, Report on the Manuscripts of the Earl of Mar and Kellie* (1904), I, 356.
56 How many saltmasters were involved is not clear. Not only did they object to the proposed ban on salt refining in Scotland, implicit in the Union settlement, but they also resented the fact that unlike English salt, the Scottish variety could not be freely used to preserve meat and fish for exportation, or for ships' stores. This ban, which went back to 1705, was still being opposed by at least three Scottish saltmasters in 1718, when they protested that the measure had been passed by the Scottish Parliament following the 'artful Insinuations and suggestions of some dealers in foreign salt', presumably about the desirability of Scottish salt for these purposes. See *Remarks for the Salt-Masters, Fishers of Salmond...upon the Eighth Article of the Treaty of Union* (1707); Journals of the House of Commons, XIX, 134. Most saltmasters, it seems, were prepared to accept the compensation of a watertight domestic market in return for the loss of whatever sales they had formerly made to the fish exporting industry.
57 Defoe, *History*, 140–1; Riley, *Union*, 290.
58 Defoe, *History*, 144; see too A Aufrere (ed), *The Lockhart Papers* (2 vols, 1817), I, 191.
59 Riley, *Union*, 276.
60 SRO, CS 96/4520, 'Cash Accompts Betwixt William Morison of Prestongrange And William Henderson his Factor Commencing March 25: 1716'.
61 Dundee Archive Centre, Customs and Excise Records, CE 53/2/2, Board to Collector (Montrose), 28 March 1733.
62 Hughes, *Studies in Administration*, 419.
63 *Ibid*, 420.
64 Wemyss Castle MSS, Ledger of Salt Accounts, 1771–6, copy letter to James Ramsay, 1 March 1771.
65 NLS, Cadell MSS, Acc 5381, Bound Volume 2, 4 December 1770, 15 March 1771.
66 Whatley, *'That Important and Necessary Article'*, 37.
67 SRO, E 536/53–5, Salt Charge Vouchers, 1767–70.
68 Wemyss Castle MSS, Ledger of Salt Accounts, 1771–6, copy letter to John Roebuck, 21 February 1773.
69 *Ibid*, copy letter to Robert Fall, 25 February 1773.
70 *Ibid.*
71 On the other hand, all of the proprietors of saltworks on the north side of the

Forth, except Inverkeithing, had been members by 1787, *Scots Magazine*, 66(1804), 817. It should be noted that by 1793, when the industry's monopoly was threatened, Sir John Hope became active within the ranks of the saltmasters.

72 Payne, 'Halbeath', 17.

73 SRO, Henderson of Fordel MSS, GD 172/503/1, James Pinkerton to Sir John Henderson, 13 March 1787.

74 Wemyss Castle MSS, Ledger of Salt Accounts, 1771–6, copy letter to William Stewart, 15 October 1775.

75 See *Scots Magazine*, 66(1804), 817; Payne, 'Halbeath', 17–18, 31–2; SRO, Henderson of Fordel MSS, GD 172/5013/1, Pinkerton to Henderson, 13 March 1787.

76 NLS, Cadell MSS, Acc 5381, 15/2, John Grieve to William Cadell, 26 December 1775.

77 See note 75 above; and for example, SRO, Henderson of Fordel MSS, GD 172/503/11, 'Account of Salt made and Sold at the Works as under in the Quarter preceding 24 July 1790'; GD 172/503/14, 'Copy Minutes of the Meeting of the Salt Proprietors at Kirkcaldy, 10 September 1790'. Members reckoned they lost 8*d*. or 10*d*. on every bushel they exported, and slightly less on each bushel of salt sold for fishery purposes.

78 SRO, Henderson of Fordel MSS, GD 172/503/3, James Pinkerton to William Wilson, 24 October 1789.

79 Derived from SRO E 536/1–84, Salt Charge Vouchers.

80 *Ibid.*

81 *Ibid.*

82 NLS, Cadell MSS, Acc 5381, 53/1, John Cadell to William Cadell, 20 September 1769.

83 SRO, Henderson of Fordel MSS, GD 172/503/11, 'Account of Salt made and Sold...Quarter preceding 24 July 1790', which shows that 4,444 bushels of salt was sent to Dumfries, from three works, Bo'ness, Grange and Cockenzie.

84 Wemyss Castle MSS, Ledger of Salt Accounts, 1771–6, copy letter to Joseph Forsyth, 24 June 1774.

85 *Scots Magazine*, 66(1804), 763, 817; at the same time salt was sold to Dumfries merchants at 1*s*. per boll.

86 *Ibid.*

87 NLS, Cadell MSS, Acc 5381, Box 15/1, John Grieve to William Cadell, 26 October 1773.

88 NLS, Cadell MSS, Acc 5381, Bound Volume 2, 9 August 1779.

89 Roger, *General View*, 16.

90 SRO, Leven and Melville MSS, GD 26/13/169, John Hopetoun to Henry Dundas, 26 October 1785; see too, for example, *Old Statistical Account*, XVII (1796), Small Isles, 291.

91 B Lenman, *Integration, Enlightenment and Industrialisation: Scotland 1746–1832* (1981), 79.

92 Invariably this duty has been seen and described as a 'burden' on the east coast coal industry. See A C O'Dell, 'A Century of Coal Transport in Scotland, 1742–1842', in L Dudley and S W Woolridge (eds), *London Essays in Geography* (1951), 239; Duckham, *Scottish Coal*, 230; Lenman, *Economic History*, 133. In fact the preferential nature of the duties had been approved of and sought by Scottish coalmasters at and in the years immediately following the Union. Proposals for change brought protests from Sir John Henderson, William

Wemyss and John Campbell WS, that this would be a 'flagrant violation of the Treaty of Union', *House of Commons Journals*, 49 (1792–3), 832.

93 SRO, Henderson of Fordel MSS, GD 172/503/20, Sir John Henderson to John Cadell, 3 December 1793.

94 *Ibid*; GD 172/496/32, 'Observations on the proposed Repeal of the Coal and Salt Tax', nd (*c*.1793).

95 Whatley, *'That Important and Necessary Article'*, 38.

96 Whatley, 'Process of Industrialisation', 166–7.

97 SRO, Henderson of Fordel MSS, GD 172/503/2, James Pinkerton to Sir John Henderson, 28 March 1788.

98 Anon (but almost certainly by John Taylor), *Considerations on the System of Laws in Scotland Relative to Salt* (1794), 24. This pamphlet can be seen in the SRO, GD 172/503/8.

99 SRO, Henderson of Fordel MSS, GD 172/503/23, John Cadell to Sir John Henderson, 30 December 1793.

100 SRO, Henderson of Fordel MSS, GD 172/503/32, 'Hints...on the Salt Laws', 10 January 1794.

101 SRO, Henderson of Fordel MSS, GD 172/503/22, William Cadell to Sir John Henderson, 30 December 1793.

102 SRO, Henderson of Fordel MSS, GD 172/503/1, James Pinkerton to Sir John Henderson, 12 March 1787.

103 SRO, Henderson of Fordel MSS, GD 172/503/34, 'Minutes of Meeting of the Proprietors of the Salt Works on the East Coast of Scotland, 19 February 1794'. The public campaign had to be stepped up: Henderson was frustrated by the lack of active support from the Scottish salt interest in Parliament: 'Most of our Salters in Parliament are either ignorant of their own Interest or too Sillily obsequious to defend it', he wrote in December 1793.

104 Hughes, *Studies in Administration*, 418.

105 *Journals of the House of Commons*, 53 (1797–8), 696–7; see also Hughes, *Studies in Administration*, 423.

106 *Hansard's Parliamentary Debates*, 3, 1805, 789.

107 Hughes, *Studies in Administration*, 418; Clows, *Chemical Revolution*, 61: SRO, Leven and Melville MSS, GD51/5/259/1, Robert Reid Cunninghame to Henry Dundas, 4 March 1799.

108 Clows, *Chemical Revolution*, 62; there were also highly effective pleas for the retention of local monopolies from English salt producers.

109 For a full account of the moves which led to the repeal of all salt duties, see Hughes, *Studies in Administration*, 357–80; 471–507.

110 H Hamilton, *An Economic History of Scotland in the Eighteenth Century* (Oxford 1963), 114–5; *Old Statistical Account*, XIII (1794), Rathven, 420; NLS, MS 640, Lord Selkirk to Henry Dundas, 11 June 1793.

111 *Old Statistical Account*, X (1794), Harris, 392; similar pleas came from many other parishes in the north and west of Scotland.

112 *Old Statistical Account*, XVII (1796), Small Isles, 291.

113 *Old Statistical Account*, XVI (1795), Whithorn, 292; See SRO, Leven and Melville MSS, GD 51/9/274/2, 'Case of the Bacon Dealers on the County of Dumfries'; GD 51/5/201, Patrick Heron to Henry Dundas, 23 September 1793.

114 Clows, *Chemical Revolution*, 62.

115 See Hughes, Studies in Administration, 424; see, for example, Report from the Committee Appointed to enquire into the Laws relating to the Salt Duties, 1805 (35), III, 76, where it was concluded that the communation of the salt duties

'would be productive of many great and important Advantages to all Descriptions of Persons in this Kingdom, and would be highly beneficial to its Agriculture, Fisheries, Trade and Manufactures.'

CHAPTER 5

1 Smout, *History of the Scottish People*, 168.
2 *Ibid*; see too B F Duckham, 'Serfdom in Eighteenth Century Scotland', *History*, LIV (1969), 178–9, although in spite of its title this essay deals solely with collier serfdom.
3 *Acts of the Parliament of Scotland*, VI, i, 761.
4 Smout, *History of the Scottish People*, 403.
5 For an example of the older view, see I A Adams, T*he Making of Urban Scotland* (1978), 53; such assumptions about the 'separateness' of the Scottish colliers are challenged by Houston, 'Marriage Formation and Domestic Industry', 225–6.
6 Beveridge, *Culross and Tulliallan*, II, 245–6.
7 Clows, *Chemical Revolution*, 56.
8 Smout, *History of the Scottish People*, 169–70.
9 *RPCS*, 2nd ser, V, 329–30.
10 NLS, Ch 649, 'Instrument of the Laird of Bogie Agt John Finlay and others, 1705'.
11 NLS, Fletcher of Saltoun MSS, MS 17679, 'Contract Between His Grace the Duke of Argyle And James Kilpatrick', 1750.
12 NLS, Ch 654, 'Obligation: Peter Pugston Salter to Sir James Wemyss of Bogie, 1757'.
13 NLS, MS 3720, Journal of the Management of the Coall and Salt Works of Prestongrange beginning from the 2nd of April 1748.
14 W Fraser (ed), *Memorials of the Family of Wemyss of Wemyss* (Edinburgh 1888), III, 203; see too J Cameron, *Prisons and Punishment in Scotland from the Middle Ages to the Present* (Edinburgh 1983), 5–7.
15 SRO, Biel MSS, GD6/1251/1, Tack, Sir Peter Wedderburn and Sir William Murray and William Nicolson, 1679.
16 NLS, Wemyss MSS, MS 5723, Tutors Sederunt Book, 7 April 1732.
17 Didsbury, 'Cheshire saltworkers', 143.
18 SRO, Rossyln MSS, GD 164/16/124/1, Agreement between Walter Smith and David Beatson, 24 February 1752.
19 NLS, MS 5723, Tutors Sederunt Book, 10 October 1730.
20 NLS, MS 3088, Bogie Salt Book, 19 February 1728.
21 NLS, MS 5723, Tutors Sederunt Book, 8 October 1720.
22 NLS, Cadell MSS, Acc 5381, Bound Volume 2, 7 June 1779.
23 Wemyss Castle MSS, Diary of Earl David.
24 See T C Smout, *A Century of the Scottish People 1830–1950* (1986), 99.
25 NLS, MS 5723, Tutors Sederunt Book, 17 October 1729.
26 Whatley, 'A Saltwork and the Community', 46.
27 NLS, MS 5723, Tutors Sederunt Book, 8 October 1731.
28 Wemyss Castle MSS, Diary of Earl David, 9 May 1659.
29 NLS, Cadell MSS, Acc 5381, Box 12/1, 'Agreement with Salters', 1 June 1801.
30 NLS, MS 3720, Inventory of Prestongrange saltwork and colliery, 27 July 1745.
31 Cadell, *Rocks of West Lothian*, 321–2.

32 'Two bed stedds made up of old timber for the Salters', for instance, are listed in SRO, Ailsa MSS, GD25/9/Box 46, 'Inventory of Materials of the Salt Pan of Turnberrie', 15 May 1733.

33 SRO, Dundonald MSS, GD 233/183/Folder 15, 'Report by Messrs Henderson, Grieve etc, as to Forest of Culross, Tullyallan Coal etc', 16 October 1795.

34 See Whatley, 'An Early Eighteenth Century Scottish Saltwork', 98; SRO, Ailsa MSS, GD 25/9/Box 46, Alexander Kinnaird to Archibald Kennedy of Culzean, 21 May 1754.

35 SRO, Fairholme Papers, CS 237/143/2/16, 'State & View of Boness Coal & Salt works before and Since 30 June 1770, the Period when Dr Roebucks Trustees accepted the Management'.

36 NLS, MS 5723, Tutors Sederunt Book, 5 October 1720.

37 NLS, Ch 651, Petition of William Foord, 11 October 1732.

38 Didsbury, 'Cheshire saltworkers', 147.

39 SRO, Hope of Craighall MSS, GD 377/361, Pinkie Pans Salt Book, 2 August 1845–26 April 1856.

40 Calculated from NLS, MS 3088, Bogie Salt Book, 10 October 1720–2 October 1732.

41 NLS, Cadell MSS, Acc 5381, Bound Volume 2, 26 June 1780.

42 For lists of their numbers and names, see, for example, SRO, CE 12/1, Salt Establishment Books, 1714–22; Signet Library, MS 106; 54, 'An Account of the Officers of the Customs and Salt Duty at the Out Ports in Scotland, their Salaries, the supposed value of each Office, their Character and by whose interest they hold their Employment, as they stood on the Establishment at the 5th January 1755'.

43 Salt officers' salaries appear to have been partly dependent upon where they were posted. Those at the main ports in 1717 were paid £6. 5*s*. per quarter; watchmen received £3. 15*s*. These rates were apparently still being paid in the 1790s, see SRO, CH 12/4, Salt Establishment Books, 1782–1798.

44 SRO, Henderson of Fordel MSS, GD 172/503/17, David Anderson to R B Henderson, 25 July 1795.

45 Kirkcaldy Customs and Excise, 42/4/4, Letter Book, Collector to Board, 21 August 1772, 11 January 1773; indicative of the authorities' zeal for economy are the events of 1745, when Jacobite troops had raided some Scottish saltworks and stolen monies collected by salt officers. Those officers who had lost sums of money to the 'Rebells' were not automatically relieved of their responsibility to reimburse the Salt Office for the missing duties. 'It must be carefully distinguished whether it was done by force or connivance of the Officers', local Collectors were instructed. SRO, E 536/31, Papers included within the bundles of Salt Charge Vouchers for Prestonpans precinct, 1745–6.

46 SRO, Assembly Papers, CH1/2/76, f 476, 'Petition of John How late of the Salt Duty', 1738.

47 SRO, Forfeited Estate Papers, FEP 1715/E 661/Box 96, Cockenzie (Winton) Saltwork Oncost Books, 1716–7, 1718–9.

48 Wemyss Castle MSS, Oncost Book, Dysart Salt, 1817–8.

49 SRO, Coal Board Records, CB 27/1, Robert Bald's Report on Preston Island Colliery and Saltwork, 1813.

50 SRO, Rosslyn MSS, GD 164/49/398, 'Acct of Smith Wadges for the pan beeting' 1715.

51 See note 46

52 See Whatley, 'An Early Eighteenth Century Scottish Saltwork', 92.

53 Hamilton Public Library, Hamilton MSS, Bundle 4, 'Note of the Carts of panwood led to the pans at Corbihall betwixt 13 May 1690 and 23 Aprill 1693 According to Adam Pearson Saltgrieve his book'.
54 Hamilton MSS, Lennoxlove, E 4/12/13, Agreement between Duke and Duchess of Hamilton and James Cornwall of Bonhard, 8 March 1681.
55 See Duckham, *Scottish Coal*, 203–6.
56 A J G Cummings, 'The York Buildings Company: A Case Study in Eighteenth Century Corporation Mismanagement', Unpublished PhD thesis, 2 vols, University of Strathclyde, 1981, II, 310; Robertson, *Scottish Railway System*, 6.
57 SRO, FEP 1715/E 661/Box 96, 'Memoriall To Sir Henry Hoghton one of the honble The Commissioners of Inquiry anent the Coal and Salt Books of Winton An:1716'; NLS, Wemyss MSS, MS 5723, Tutors Sederunt Book, 7 October 1721, 8 April 1722.
58 *Ibid.*
59 Wemyss Castle MSS, Extracts from Wemyss Castle Muniments, III, 'Discharge—Doctor Hay for Several Wemyss Salters Fines £108 Scots, 27 March 1753'.
60 Duckham, *Scottish Coal*, 243
61 Hamilton MSS, Lennoxlove, F1/727, 'Instructions from her Grace the Duchess of Hamilton to John Hamilton, Baillie and Chamberland of Arran', 11 March 1700.
62 W R Scott (ed), *The Records of a Scottish Cloth Manufactory at New Mills, Haddingtonshire 1681–1703* (Scottish History Society 1905), xiv, 37.
63 S Pollard, *The Genesis of Modern Management* (1965), 218–9.
64 NLS, Wemyss MSS, MS 5723, Tutors Sederunt Book, 8 October 1725.
65 R S Fitton and A P Wadsworth, *The Strutts and the Arkwrights 1758–1830* (Manchester 1958), 236.
66 Pollard, *Modern Management*, 190–1.
67 T C Smout, ' Lead-mining in Scotland, 1650–1850', in P L Payne (ed), *Studies in Scottish Business History* (1967), 123.
68 R A Houston, 'Coal, class and culture: labour relations in a Scottish mining community, 1650–1750', *Social History*, 8 (1983), 2.
69 See E P Thompson, 'Patrician society, plebeian culture', *Journal of Social History*, 7 (1974), 382–405.
70 See E P Thompson, 'The Moral Economy of the English Crowd in the Eighteenth Century', *Past and Present*, 50 (1971), 78–9; for meal rioting in Scotland, see K J Logue, *Popular Disturbances in Scotland 1780–1815* (Edinburgh 1979), 18–53.
71 Smout, 'Scottish Landowners', 218.
72 Wemyss Castle MSS, Diary of Earl David, 1 October 1675.
73 See, for example, NLS, MS 3088, Bogie Salt Book, 1720–32; of course the salters were not alone in this. See Duckham, *Scottish Coal*, 289–91.
74 For example, NLS, Wemyss MSS, MS 5723, Tutors Sederunt Book, 6 April 1733.
75 Cummings, 'York Buildings Company', II, 324–5.
76 Ramsay of Wardlaw MSS, 56, James Watt to Sir John Ramsay(?), 22 March 1689.
77 Hamilton MSS, Lennoxlove, F1/839, 'Ane Scroll of His Grace the Duke of Hamilton's Coall and salt work In the Cock on Aron from the 1 to 29 day of July 1711'.
78 NLS, Wemyss MSS, MS 5723, Tutors Sederunt Book, 21 May 1725.

79 Hamilton MSS, Lennoxlove, F1/818/5, 'Expenses of ye Salters that went to Saltcoats to try the coall in Arran how it made salt', September 1708.
80 NLS, Adv MS 80.1.3, Dundas of Dundas, Patrick Maxton to George Dundas, 26 January 1727.
81 SRO, Rosslyn MSS, GD 164/49/398/10, 'Accott of Money Depursed...St Clair Salt Works Since 25 July 1713'; and see Wemyss Castle MSS, Diary of Earl David, 1 October 1675, where he noted, 'I have nothing but this large firlot to uphold my charges', after recording that he had no iron bands on the wooden firlots as 'when daily or hourly...the salters deliver the salt with them it is to my prejudice'.
82 Whatley, 'A Saltwork and the Community', 52–3.
83 SRO, Carron Company Records, GD 58/16/1, 'Coal and Salt, Weekly Checks and Oncosts', 1760.
84 Whatley, 'A Saltwork and the Community', 53, 56.
85 Wemyss Castle MSS, Ledger of Salt Accounts, 7 February 1777.
86 NLS, Cadell MSS, Acc 5381, Bound Volume 2, 22 February 1779.
87 Duckham, *Scottish Coal*, 263; 'Serfdom', 188–9; Lenman, *Economic History*, 132.
88 Tucker, 'Report', 5.
89 Clows, *Chemical Revolution*, 55.
90 C Hill, 'Pottage for Freeborn Englishmen: attitudes to wage labour in the sixteenth and seventeenth centuries', in C H Feinstein (ed), *Socialism, Capitalism and Economic Growth* (Cambridge 1967), 339–40.
91 J Rule, *The Labouring Classes in Early Industrial England 1750–1850* (1986), 107.
92 8 Geo II c.12 1735.
93 Tucker, 'Report', 6–7.
94 Wemyss Castle MSS, Diary of Earl David, 6 May 1659; at Torry the salters had to deliver 14 firlots of 'hot salt' for 12 'karts' of coals, SRO, RH9/1/35, Accounts of Torry Coal and Salt Works, 1679–80.
95 SRO, Rosslyn MSS, GD 164/Box 49/398, 'The Accompt of Expenses made of the Coal and Salt works of Torry Betwixt the 6th and 13th dayes of January 1678.'
96 SRO, Rosslyn MSS, GD 164/Box 11/6/3, Accounts, Elphinstone pans, 1716–9.
97 SRO, Forfeited Estates Papers, FEP 1715, E 661/Box 196, 'Salt Book of Cockenzie from 1 July–July 1717, keeped by Mr William Adam, Saltgrieve there.'
98 Pollard, *Modern Management*, 46, 216, 255–6; J Rule, *The Experience of Labour in Eighteenth Century Industry* (1981), 124–34.
99 See, for example, NLS, Adv MS 80.1.3, Dundas of Dundas, Patrick Maxton to George Dundas, 20 April 1727; MS 3088, Bogie Salt Book, 4 May 1724.
100 See, for example, the records of Grange coal and saltwork in the 1760s and 1770s, especially NLS, Acc 5381, Bound Volume 2, Memorandum Book, 1771–9.
101 SRO, E 536/9, 'Account of Fines actually levied by John Haldane, Collector, 3 November 1715–29 September 1724' (Prestonpans precinct); CS 96/4520, 'Cash Accompts...William Morison of Prestongrange And William Henderson his factor, 25 march 1716–February 25 1725'.
102 *Calendar of Treasury Papers*, CCXLVII, 275.
103 SRO, CE 8/5, Letter book, Scottish Excise Board to Treasury, 9 April 1798.
104 See Whatley, *'That Important and Necessary Article'*, 54–5; SRO, CE 23/4/1792,

Letter book, Board to Collector, 5 February 1795; CE8/7, Letter book, Scotish Excise Board to Treasury, 15 January 1805.

105 Wemyss Castle MSS, Ledger of Salt Accounts, 2 March 1773.

106 J Jack, *An Historical Account of St Monance, Fifeshire* (Edinburgh 1844), 147.

107 SRO, Henderson of Fordel MSS, GD 172/503/17, David Anderson to R B Henderson, 25 July 1793.

108 Kirkcaldy Customs and Excise, Excise General Letter Book, 63/6/4, 1 September 1808.

109 Stevenson, *Presbytrie Book*, 242; Wemyss Castle MSS, Diary of Earl David, 3 December 1660.

110 SRO, Henderson of Fordel MSS, GD 172/503/43, General letter to all Collectors from the Excise Office, Edinburgh, 1 September 1808.

111 NLS, Cadell MSS, Acc 5381, Bound Volume 2, 5 January 1779.

112 Wemyss Castle MSS, Diary of Earl David, 17 January 1666.

113 NLS, Pitfirrane MSS, MS 6432, f 155, 'Accompt of work to Sir Peter Halkett', 1712–3.

114 SRO, Hope of Craighall MSS, GD 377/361, Pinkie Pans Salt Book, 1845–56.

115 SRO, Ailsa MSS, GD 25/9/Box 46, 'Declaration upon visitation of the houses [and] Salt pan of Turnberry, 27 May 1754'; SRO, CS 229/P1/30, List of Feuars of Prestongrange.

116 Calculated from SRO, Forfeited Estates Papers, FEP 1715/Box 196, Weekly Output Books, 1716–8.

117 Wemyss Castle MSS, Diary of Earl David, 15 May 1654; NLS, Cadell MSS, Acc 5381, Bound Volume 2, 22 February 1779.

118 Wemyss Castle MSS, Ledger of Salt Accounts, 24 February 1772.

119 NLS, Cadell MSS, Acc 5381, Bound Volume 2, 26 June 1780.

120 *Ibid*, Box 39/I, Abstracts, Maryburgh and Craigie saltworks, 1777; SRO, CS 228/O/1/41, W Orr v Earl of Eglinton, 'Condescendence for Wm Orr & S Mitchell, pursuers v Alexander, Earl of Eglinton, Charles Sheriff & John Fowlis, defenders', 1764.

121 Wemyss Castle MSS, Diary of Earl David, 3 December 1660; in addition to their money incomes, salters also received rent-free housing and generally, a supply of coals.

122 Duckham, *Scottish Coal*, 263, quotes Bald's figure for collier earnings of 1*s*. 2*d*. per day. On the optimistic assumption that they worked five days a week, fifty weeks of the year, their earnings would have been in the region of £15. A daily hewing rate of this order is also given in Flinn, *Coal Industry*, 387; colliers at the Marquess of Lothian's pits in the 1740s came closest to the salters' figures, and could apparently reasonably earn 10*s*. 2*d*. per week, NLS MS 2980, f 28, 'Memorandum To the Most Honourable The Marquess of Lothian Anent His Colliers Wages, 1744'.

123 Hamilton MSS, Lennoxlove, F1/839, 'Ane Scroll of His Grace the Duke of Hamilton's Coall and Salt work on the Cock of Aron from the 29 july to 26 August 1711'.

124 *Ibid.*

125 For example, SRO, Cardross Writs, GD 15/903/1–60, Kincardine coal and salt accounts, 1740; Ailsa MSS, GD 25/9/Box 46, Accounts, Turnberry saltwork, 1754; CS 228/O/1/41, Orr v Eglinton, 'Condescendence', 1764.

126 Duckham, *Scottish Coal*, 263–4.

127 Calculated from NLS, Cadell MSS, Acc 5381, Box 39/I, Abstracts, Maryburgh and Craigie saltworks, 1777.

128 Calculated from SRO, Hope of Craighall MSS, Pinkie Pans Book, 4 November 1797–3 November 1798; see too *Old Statistical Account*, II, Prestonpans, 570.
129 Duckham, *Scottish Coal*, 271–4.
130 For a recent survey of wage rates which covers this period, see P H Lindert and J G Williamson, 'English Workers' Living Standards during the Industrial Revolution: A New Look', *Economic History Review*, XXXVI (1983), 4.
131 Girvin, *Impolicy*, 38.
132 NLS, Cadell MSS, Acc 5381, Box12/1, 'Agreement with Salters', 1 June 1801.
133 SRO, Rosslyn MSS, GD 164/45/348, 'Cost of Salt Making at Dysart', 23 October 1816.
134 Wemyss Castle MSS, Oncost Book, Dysart Salt, 1817–8.
135 Calculated from SRO, Hope of Craighall MSS, GD 377/361, Pinkie Pans Salt Book, 1845–56.
136 Anon, *Riot at Musselburgh. A Full and particular account of a...Riot...which took place on Musselburgh Links yesterday, Thursday 21st of July, 1823, between the Irish, and Colliers, and Salters etc.*'
137 Lenman, *Economic History*, 125.

CHAPTER 6

1 See, for example, T S Ashton and J Sykes, *The Coal Industry of the Eighteenth Century* (Manchester 1929), 78–9; Smout, *History of the Scottish People*, 405–6; Duckham, *Scottish Coal*, 294-301; Ferguson, *Scotland*, 188–9.
2 Duckham, 'Serfdom', 178.
3 Stair, *Institutions*, 1013.
4 Ashton and Sykes, *Coal Industry*, 78.
5 Duckham, 'Serfdom', 193; R H Campbell, *Scotland Since 1707: The Rise of an Industrial Society* (Oxford 1965), 128; Lenman, *Economic History*, 132; Butt, *Industrial Archaeology*, 133.
6 Campbell, *Lanarkshire Miners*, 16.
7 Lenman, *Economic History*, 125.
8 Hughes, *Studies in Administration*, 413; *Scots Magazine*, 66 (1804), 760.
9 Of these, 188 were men, 12 were youths and five were women; for an estimate of the total number of workers employed directly at Scotland's saltworks in the early eighteenth century, see Whatley, 'A Saltwork and the Community', 57–8.
10 SRO, Fairholme Papers, GD 237/143/2/16, 'State & View of Boness Coal & Salt works before & Since 30 June 1770'; CB 27/1, Robert Bald's Report of Preston Island Colliery and Saltworks, 1813.
11 Nef, *British Coal Industry*, II, 145; Flinn, *Coal Industry*, 339, 345–9.
12 SRO, CS 238/5/S/10/9, W Scott v J Dewar, 1781, Copy letter, Walter Scott to Lord Advocate, 18 March 1774; no case of a salter becoming a collier has yet been discovered.
13 *Ibid.*
14 Late in the summer of 1774, for instance, William Cadell wrote that colliers were 'plentier' than they had been for many years, NLS, Cadell MSS, Acc 5381, 53/1, William Cadell to John Cadell, 10 August 1774.
15 SRO, Fairholme Papers, GD 237/143/2/16, 'State & View of Boness Coal & Salt works'.
16 Derived from SRO, E 536, Salt Charge Vouchers; see too Whatley, 'Sales of Scottish Marine Salt', 12; the Prestonpans figures are given in Lenman, *Economic History*, 125.

17 SRO, CS 238/5/S/10/9, Scott v Dewar, Patrick Robertson to Walter Scott, 13 April 1774.
18 Duckham, 'Serfdom', 194–5.
19 Whatley, 'From Servitude to Subjugation', 15.
20 Duckham, *Scottish Coal*, 249–50.
21 See note 15 above.
22 Smout, *History of the Scottish People*, 169; G Donaldson, *Scotland: The Shaping of a Nation* (1980 ed), 232.
23 Act 39 Geo.III c.56.
24 Duckham, *Scottish Coal*, 306–7; Campbell, *Lanarkshire Miners*, 16-7.
25 *Ibid*, 93–100.

INDEX

(Compiled by H A Whatley)